INVERSIVE GEOMETRY

INVERSIVE GEOMETRY

FRANK MORLEY

AND

F. V. MORLEY

DOVER PUBLICATIONS, INC.

MINEOLA, NEW YORK

TO

L. J. M.

Bibliographical Note

This Dover edition, first published in 2014, is an unabridged republication of the work originally published by Ginn and Company, Boston, in 1933.

Library of Congress Cataloging-in-Publication Data

Morley, Frank, 1860–1937.
Inversive geometry / Frank Morley, F. V. Morley.
pages cm.
Originally published: Boston : Ginn and Company, 1933.
Includes index.
ISBN-13: 978-0-486-49339-8
ISBN-10: 0-486-49339-3
1. Geometry, Modern. 2. Inversions (Geometry) 3. Geometry, Algebraic. I. Morley, F. V. (Frank Vigor), 1899–1980. II. Title.

QA473.M7 2014
512'.1—dc23

2013025733

Manufactured in the United States by Courier Corporation
49339301 2014
www.doverpublications.com

PREFACE

This book is an introduction to algebraic geometry, with special reference to the operation of inversion. It originated in the effort to show the use of modern views in comparatively elementary and practical questions; and thus to afford both a review of these questions and a suitable preliminary to many specialized courses.

The algebra is ordinary algebra, where one adds and multiplies numbers. At the outset one should avoid any limitation of these numbers. They are represented as the points of a plane. In this plane is a line on which are represented the so-called real numbers.

The geometry is in the beginning Euclidean, and throughout we have a Euclidean background. This conception is at once so familiar and so convenient that it is essential for the purpose of an introduction to spatial universes.

In Part I we go over the application of numbers to Euclid's planar geometry, and arrive at the Euclidean group of operations. By adjoining an inversion we get the wider inversive group. We thus get the idea that by isolating a point of a sphere (which point we call infinity) we have the Euclidean case of the plane. But also we get the idea that it would be convenient to isolate, say, a circle of the sphere. The geometry is now non-Euclidean; it has a non-degenerate infinity, and, as infinity affords the best reference-scheme, it is a better geometry in general. So for the consideration of the space around one it is convenient to regard it as bounded by a sphere. This way of introducing Cayley's absolute seems pedagogically entirely proper.

So far we have sought to apply Klein's views of a geometry to the inversive group. A chapter follows on flows of the simplest kind, which may serve to introduce the subject of conformal or isogonal mapping, and to emphasize the fact that the study of an analytic function is the study of a region. And a chapter is added on Differential Inversive Geometry, with reference to the recent work of Pick, Liebmann, Kubota and others.

The chapters in Part II are mainly of the nature of applications. Some are quite elementary; for instance if one seeks invariants or covariants under the Euclidean group of a set of points or lines it is proper to begin with the triangle.

We believe that the tradition that simple geometric and mechanical questions are to be handled only as Euclid or Descartes might have handled them is very hampering; that the ideas of Riemann, Poincaré, Klein, and others have pleasant reverberations in the investigation of elementary questions by students of proper maturity and leisure.

These applications may well be studied first, and the theoretic chapters referred to as occasion requires.

It is a pleasant duty to express thanks to the University of California where in a course of summer lectures the plan of this book originated; to The Johns Hopkins University where the analytic geometry, to which this aims to be an introduction, could be further discussed; and to the Carnegie Institution of Washington for its benevolent interest in the publication. We are also grateful for the courtesy and care of the publishers and printers.

FRANK MORLEY
M.A., Sc.D. (Cambridge)
Professor-Emeritus of the Johns Hopkins University
Research Associate of the Carnegie Institution

F. V. MORLEY
D.Phil. (Oxford)

CONTENTS

PART II

PART I

CHAPTER I

OPERATIONS OF ELEMENTARY GEOMETRY

§ 1. **Instruments** – Euclid's plane geometry deals with figures, such as triangles, parallelograms, circles, in a plane. These are not pre-existent; they have to be either drawn or printed. In the phrase of Enriques,* they are *werdende*, not *fertige*. The drawing is done by instruments such as a straight rod or rule, or a combination of rods such as a triangle or a pair of compasses.

Let the instrument be a card lying on a table. We can then (thanks to the axioms) move the card freely over the table.

The card has a boundary; we take as the boundary a triangle, with three straight edges 1, 2, 3, and with the upper face marked. What we understand by straight is that the card can be turned over, keeping, say, the edge 1 fixed. The card will again fit the table, if both card and table are flat, and if the edge is straight. The marked face is now down.

We take this overturn as one of the fundamental operations of Euclidean geometry. If at first a point marked b of the card is over a point marked a of the table, then after the overturn the point marked b is over another point, say, $\bar{a}$ of the table. There is then set up in the table a correspondence of points with a line of fixed points, which we call a reflexion. In the table is the record of the operation. By a usual trick of language, the same word reflexion will serve both for the operation and the correspondence which it sets up. Thus, instead of overturn we write reflexion.

Let us now make a second reflexion or overturn of the card, say about the edge marked 2. The marked face is now up. It seems that the result of the two reflexions could

* *Encyklopädie der Mathematischen Wissenschaften*, vol. 3, 1 Teil, Seite 6.

be obtained by moving the card along the table. Thus we get the idea that a planar displacement can be analysed into two reflexions. And this we take as a definition: a planar displacement is the product of two reflexions.

When we study Euclid from the point of view of the operations involved, it is essential to distinguish a point of the table (which is given or fixed) from the coincident point of the card. It is convenient to call the latter (fixed in the card) a particle.

It will be understood that a figure in a table or a card on a table is merely suggestive. We need not act—we can imagine. And the imagined figure or operation is accurate. When we have introduced algebra, the appropriate symbolism or shorthand, we have the other extreme, the manipulation of symbols. Between the two—the material and the symbolism—lies the geometry.

And so for a space, a solid body is suggestive only. The idea of the Euclidean space and the solid body lying on it (not in it) is accurate, both for the space and the body. To connect the body and the idea of it requires an axiom stated by Einstein * thus: "Solid bodies are related with respect to their possible dispositions, as are bodies in Euclidean geometry of three dimensions."

A second experiment is to fold a sheet of paper. Mark a point a on the paper, and fold the paper about a line not on the point. After folding the point is marked on both sheets; we have coincidence of two points a and $\bar{a}$. When the paper is unfolded we have a and $\bar{a}$ as images on the line of the fold. The correspondence thus set up is again a reflexion.

A reflexion is then a mutual or involutory correspondence, said to be of period 2 because when repeated it gives identity. It is convenient to state it by an axiom. Thus, after Hjelmslev,† we state the axiom of reflexion:

Among one-to-one mutual correspondences of the points of a plane there is one and only one other than identity which

* *Side-lights on Relativity*, London, Methuen, p. 32.

† *Math. Annalen*, vol. 64. See also Pascal, *Repertorium der Höheren Mathematik*, Teubner, 1910, vol. 2 ; and Schur, *Grundlagen der Geometrie*.

leaves all points of a line l fixed; which leaves no other points fixed; which assigns to any line l_1 a line $\bar{l}_1$, and which assigns to any point x of l_1 a point $\bar{x}$ of $\bar{l}_1$.

This is necessary for the existence of a line. It is in no way sufficient, for it is true of circles. For a suitable form of the other axioms, when reflexion is made primary, we refer to Hjelmslev.

§ 2. **Rotations** – The line and the reflexion are so interwoven that it is convenient here to use the same letter L for both. The context shows which is meant.

A point a under a reflexion L_1 becomes aL_1; under two reflexions L_1, L_2 in this order it becomes aL_1L_2. A reflexion effected twice gives identity, I. Thus $aL^2=a$, or simply $L^2=I$.

Consider two reflexions, in an assigned order, about lines which meet at c. Their product is the operation known as a rotation. The point c is the fixed point, or centre. As a continuous motion the card rotates about the point c, through an angle which is twice the angle which L_2 makes with L_1. The motion is rigid. A point of the card describes an arc of a circle. Any integer number of complete revolutions may be included. They do not affect the displacement.

In general L_2L_1 is the reciprocal or undoing of L_1L_2, since $L_1L_2L_2L_1=I$. Identity for rotation is any number of revolutions.

In general, then, L_1L_2 is not L_2L_1. When they are the same we have a semi-revolution or reversion. The lines L_1 and L_2 are then perpendicular; or at right angles, or orthogonal, or normal. The reflexions are also said to be orthogonal or normal.

When $L_1L_2=L_2L_1$, then $L_1L_2L_1L_2=I$. This is written $(L_1L_2)^2=I$.

Similarly we have $(L_1L_2)^3=I$ when the two lines make an angle $\pm\pi/3$. And $(L_1L_2)^n=I$ when the angle is π/n.

Any point other than c gives under the operations $(L_1L_2)^{\pm n}$ a sequence of points which is a special geometric progression; it is an unclosed regular figure which we may call an elliptic scale. When for a given n and no smaller integer $(L_1L_2)^n=I$, the sequence is closed and is the vertices of a regular polygon, convex or crossed.

If four lines L_i ($i=1$, 2, 3 or 4) form a rectangle, we have

$$(L_1L_2)^2=(L_2L_3)^2=(L_3L_4)^2=(L_4L_1)^2=I.$$

§ 3. **Translations and Reversions** – When we take two parallel lines we call the operation L_1L_2 a translation, T. It is a rigid displacement of the card over the table. If it carries a point of the card (or particle) from the position 0 of the table to the position 1, then it carries the particle originally at 1 to a position marked 2. Thus T^2 carries the particle at 0 to 2, and so on. The repetitions of T give an arithmetical progression of points.

The operation L_2L_1 is T^{-1}, the reciprocal or undoing of T. It carries the particle at 0 to the position -1. Generally $T^{\pm n}$, where n is a natural number or positive integer, gives the complete arithmetical progression, or linear scale.

When we have two perpendicular lines L_1 and L_2 we call the operation L_1L_2 (or L_2L_1) a semi-revolution (of the card) or a reversion (in the table), about the point where the lines meet.

The point and the reversion are so interwoven that it is convenient for the moment to denote either by the same letter P.

We note that $P^2=I$. Thus operations which, done twice, give identity—the square roots of identity—are, so far, I itself, any reflexion, or any reversion.

Exercise 1 – The operations of Euclid which leave the figure of two points invariant form a group of four operations, namely, reflexion L_1 in the join, reflexion L_2 in the axis, L_1L_2, and I.

Under a reversion P, a line L becomes a parallel line or is fixed. For if L and LP meet, say at a, they meet also at aP.

Two points a, b and the points aP, bP are successive vertices of a parallelogram. Conversely a parallelogram is a figure (a special quadrangle) which admits an operation P, or is invariant under P.

A reversion about the base-point 0 is denoted by the minus sign. Thus a and $-a$ are reverse points, as to 0.

§ 4. **The Product of Reversions** – To compound P_1 and P_2 we consider the join L of the two points and the lines L_1 and L_2 perpendicular to L at the points. Then

$$P_1 = L_1L \quad \text{and} \quad P_2 = LL_2$$

Hence

$$P_1P_2 = L_1LLL_2 = L_1L_2$$

The product is then the product of reflexions in two parallel lines, that is a translation.

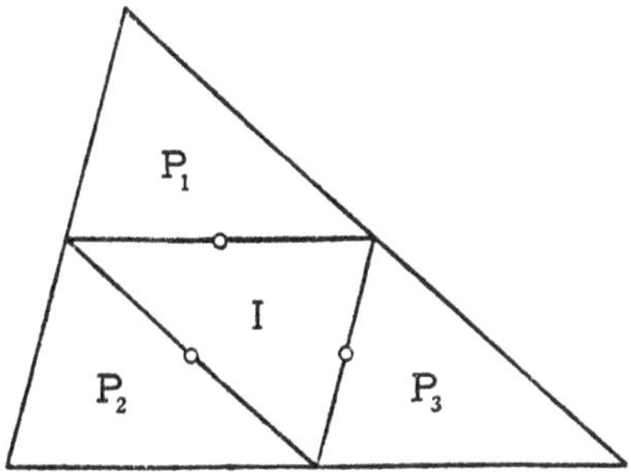

FIG. 1

Thus a translation can be broken up into two reversions, one of which is arbitrary. If we have two translations T_1, T_2, then we can write

$$T_1 = P_2P_3 \qquad T_2 = P_3P_1$$

whence

$$T_1T_2 = P_2P_1$$

That is, the result of two translations is a translation. Or symmetrically if we take three reversions P_i then P_2P_3, P_3P_1, P_1P_2 are three translations whose result is identity. This is illustrated in fig. 1. There is a triangle and the mid-points P_i of the constituent edges. Marking the inner triangular region as I (identity), the outer triangular regions will be P_i. P_1P_2 sends region P_1 into I, and I into region P_2, that is P_1 into P_2. This is a translation and the product $P_1P_2 \cdot P_2P_3 \cdot P_3P_1$ is identity.

A translation as applied to a given point is called a vector. Mark this point 0. Then, if the translation T_1 brings the particle at 0 to the point a, and T_2 brings it to the point b, then T_1T_2 brings it to the point c, where 0, a, c, b is a parallelogram (fig. 2). This is the parallelogram law, a definition of addition for vectors. We write

$$c = a + b$$

It is to be emphasised that a and b and c are not yet

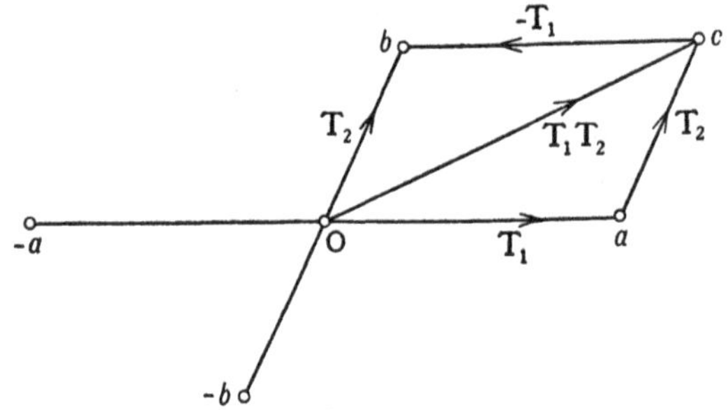

FIG. 2

numbers. To have a number we must both add and multiply. We must have what are called in algebra terms and factors.

The vector is a term, and two vectors can only be added. It is better then to speak of the addition of points. We add two points by the parallelogram law. The addition is commutative—that is,

$$a + b = b + a$$

Fig. 3 serves to show that this addition of points is also associative—that is, that

$$(a + b) + c = a + (b + c) = a + b + c$$

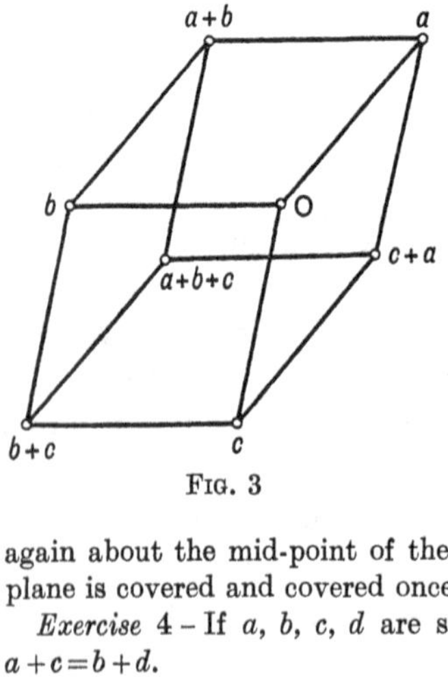

FIG. 3

It suggests a parallelepiped; and it is to be noted that this process of addition is applicable to a Euclidean space, or to any Euclidean spread.

Exercise 2 – If for four reversions $P_1P_2P_3P_4 = I$, show that the four points form a parallelogram.

Exercise 3 – A quadrangle is reversed about the mid-point of each edge, and again about the mid-point of the opposite edge, and so on. Show that the plane is covered and covered once.

Exercise 4 – If a, b, c, d are successive vertices of a parallelogram, then $a + c = b + d$.

We have here the equal vectors $b-a$ from a to b, and $c-d$ from d to c; or $d-a$ and $c-b$.*

§ 5. **Stretches** – So far operations have conserved both shape and size. There is next the case when the size is altered, as in Euclid's theory of similar triangles.

In arithmetic, viewed geometrically, one takes a rectangular card and marks the line midway between two opposite ends, for instance, by folding so as to bring the ends together. Marking the ends as 0 and 1, the mid-line is $\frac{1}{2}$. Of the two halves we again mark the mid-lines, for instance by folding twice. These mid-lines are $\frac{1}{4}$ and $\frac{3}{4}$.

The binary notation is here convenient. In the scale of 2, the line $\frac{1}{2}$ is ·1, the line $\frac{1}{4}$ is ·01, the line $\frac{3}{4}$ is ·11. A third halving of the four quarters gives the new lines $\frac{1}{8}$, $\frac{3}{8}$, $\frac{5}{8}$, and $\frac{7}{8}$, or in the binary notation ·001, ·011, ·101, and ·111. We assume that by continuing this process we can distinguish between a proposed line and any other line. That is, that any line parallel to the ends can be named by the series

$$a_1/2 + a_2/2^2 + a_3/2^3 + \; . \; . \; .$$

where a_i is either 0 or 1.

For the ends themselves, every a_i is 0 in the one case, and 1 in the other.

Having so constructed a ruler, to the approximation desired, we translate it over the plane, end to end. We have then on the plane an infinite strip, with not only the lines $0, \pm 1, \pm 2 \; . \; . \; .$ marked, but all intervening lines.

Two points, 0 (zero) and 1 (unity), being then taken in a plane, we assign to each point of the ray from 0 to 1 a positive number

$$a_0 + a_1/2 + a_2/2^2 + \; . \; . \; .$$

where a_0 is an integer n, including 0, and to each point of the complementary ray from 0 to -1 a negative number, the same expression where a_0 is a negative integer.

We call the first ray the base-ray, and the positive number the distance from 0 to the point. The second ray is the

* There is need of a notation for the line on two points a and b. We shall write, on occasion, the line $a-b$, or $b-a$.

negative-ray, and the negative number is the negative distance.

The whole line, in the sense from 0 to 1, is the base-line, or the axis of real numbers. Positive and negative numbers are real numbers or simply reals. This measuring—that is, the process of naming the points of a line by a real number—depends on the repeating of an operation, or doing it n times, and halving it or doing it half a time.

Another important case of measuring is to name an angle at a point 0. In this case we have a natural unit—the right angle. To this we can assign any real number we please; the convenient number is the transcendental number $\pi/2$. To name an angle "in the first quadrant" from the ray 0 to the ray $\pi/2$ we bisect the right angle from 0 to $\pi/2$ internally, getting the mid-ray $\pi/4$. Of the two halves we again mark the mid-rays, which are $\pi/8$ and $3\pi/8$. We assume that by continuing this process we can distinguish between any proposed ray and any other ray, both in the first quadrant. That is to say, to any such ray can be attached a positive number:

$$\theta = a_1\pi/4 + a_2\pi/8 + a_3\pi/16 + \ . \ . \ .$$

where each a_i is either 0 or 1.

This number is the acute angle.

For any ray we may add the right angle $\pi/2$ or subtract one or two right angles. We have then a positive angle, for which θ is between 0 and π, and a negative angle, for which θ is between $-\pi$ and 0. The principal angle is θ, where $-\pi < \theta \leqslant \pi$. The rotation as a displacement is the same for θ and for $\theta \pm 2n\pi$. As a continuous motion, the actual amount of turning is of course essential.

By adding the numbers we add the angles. But the sum of principal angles may not be a principal angle.

Exercise 5 - Apply the formula $\frac{1}{3} = \frac{1}{2} - \frac{1}{4} + \frac{1}{8}$ . . . to the trisection of an angle.

If now we take the ray from 0 to any point a, we can name any point on this ray by ap, where p is a positive number.

And taking two points a and b, we take ap on the ray to a, bp on the ray to b. Euclid's theory states that $a-b$ and $ap-bp$ are parallel.

If we join (fig. 4) b to ap, the parallel on bp gives ap^2. If we join a to bp, the parallel on b gives a/p or ap^{-1}. We construct thus on either ray a geometrical progression with positive ratio or radial scale. The point 0 is inaccessible.

Equally if we take the line on 0 and a, any point on it is $a\rho$ where ρ is a real. And taking a second line on 0 and b, the join of $a\rho$ and $b\rho$ is parallel to the join of a and b.

We have then for a given real ρ a one-to-one correspondence in the plane: to any point x there is a point ρx. This is (using Clifford's word) a stretch, about the fixed point 0.

A convenient instrument is the pantograph.

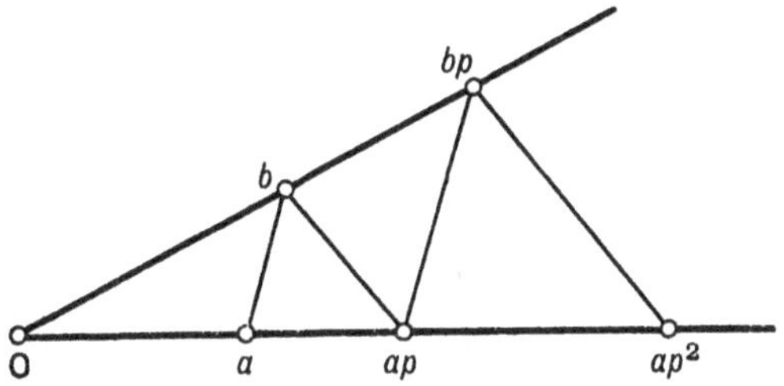

FIG. 4

As effected by a continuous operation a stretch is a case of flow (Chapter X).

A stretch sends a triangle into a positively similar triangle. Corresponding edges are parallel. They are positively parallel when ρ is positive, and negatively parallel when ρ is negative.

A stretch about 0 is then multiplication of a point by a real number. It is important to notice that two stretches about 0 are commutative—that is, that $\rho\rho'$ is $\rho'\rho$.

An elegant geometric proof is supplied by the theorem of Pappus.*

Let two circles meet at a and d, and let a line on a meet them at b and c, and a line on d meet them at c' and b'. Then in all cases

$$< bad = < bc'd, \qquad \text{mod } \pi$$
$$< dac = < db'c, \qquad \text{mod } \pi$$

* G. Hessenberg, "Begründung der elliptischen Geometrie," *Math. Ann.*, vol. 61 (1905), p. 173 ; H. G. Forder, *Euclidean Geometry*, p. 154.

Whence

$$< bc'd + \angle\, db'c = \pi, \qquad \text{mod } 2\pi$$

that is, b, c' is parallel to b', c in all cases.

Adjoin to the figure the circle on d, b, c, meeting the line $dc'b'$ again at a' (fig. 5).

Then

$$b,\ c' \parallel c,\ b'$$
$$c,\ a' \parallel c',\ a$$
$$a,\ b' \parallel a',\ b$$

This is the theorem of Pappus: that when we have points a, b, c and a', b', c' on two lines, if b, c' is parallel to c, b' and

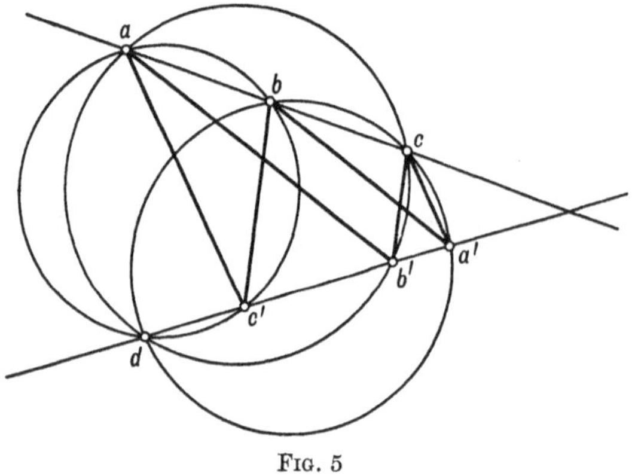

FIG. 5

c, a' is parallel to c', a, then a, b' is parallel to $a'b$. That is, there exist hexagons a, b', c, a', b, c' whose opposite sides are parallel and whose alternate vertices lie on two lines.

Applied to the present case, the theorem says that $a\rho'\rho - b\rho\rho'$ is parallel to $a - b$; so that the points $\rho'\rho$ and $\rho\rho'$ are one point.

§ 6. **Parallel Co-ordinates** – By adding the point λa to the point μb, where λ and μ are real, we get all points of the plane. This is the method of parallel co-ordinates; λ and μ being the real co-ordinates. Similarly for a space, given the points 0, a, b, c, any point is $a\lambda + b\mu + c\nu$, where λ, μ, ν are real.

Exercise 6 – The point $(a\lambda + b\mu)/(\lambda + \mu)$ is on the line joining a and b.

Exercise 7 – Three points a, b, c are on a line if

$$a\lambda + b\mu + c\nu = 0$$

where

$$\lambda + \mu + \nu = 0$$

In the plane, any point c is of the form $\lambda a + \mu b$—that is, there is between three points a relation which we write homogeneously as $\lambda a + \mu b + \nu c = 0$.

The real numbers λ, μ, ν are the barycentric co-ordinates of 0 with regard to the triangle abc. Applying a translation so that 0 is now d, we have

$$\lambda(a - d) + \mu(b - d) + \nu(c - d) = 0$$

or

$$\lambda a + \mu b + \nu c + \rho d = 0$$

where

$$\lambda + \mu + \nu + \rho = 0$$

Thus, for any point x,

$$\lambda(x - a) + \mu(x - b) + \nu(x - c) + \rho(x - d) = 0$$

The real numbers λ, μ, ν, ρ are proportional to the areas of the triangles abc, dca, dab, cba, or of bcd, $-cda$, dab, $-abc$. But we cannot as yet express an area.

CHAPTER II

ALGEBRA

§ 7. **Algebra** – We consider the application of ordinary algebra to the geometry of the plane.

Algebra is the study of a class of symbols called numbers, which under two rules of combination called addition and multiplication give a number.

The axioms of addition are:

(1) Transitive, $a+b=c$

(2) Associative, $(a+b)+c=a+(b+c)=a+b+c$

(3) Commutative, $a+b=b+a$

There is in particular a number 0 (zero) for which

$$a+0=0+a=a$$

The axioms of multiplication are the same:

$$a\times b=c \tag{4}$$

$$(a\times b)\times c=a\times(b\times c)=a\times b\times c \tag{5}$$

$$a\times b=b\times a \tag{6}$$

There is in particular a number 1 (the unit) for which

$$1\times a=a\times 1=a$$

There is then the axiom of distribution:

$$a\times(b+c)=(a\times b)+(a\times c) \tag{7}$$

We may think of a particular number; it is then given or a constant. Or we may think of a number as any number; it is then a variable. Usually, in accord with tradition, we use the early letters a, b, c . . . for constants, the later letters . . . x, y, z for variables.

Thus, if we write instead of (1),

$$y=x+b \tag{8}$$

we think of x as a variable, and y as a second variable, depending on or a function of x. Addition has thus become an operation or transformation, say T, performed on x.

Here $xT = x + b$, $xT^2 = x + 2b$, and so on. And, if we write instead of (4),

(9) $$y = ax$$

we have another operation T, such that $xT = ax$, $xT^2 = a^2x$, and so on. And, if

(10) $$y = ax + b$$

we have a more general operation, with two constants a and b.

Here $$xT = ax + b$$
$$xT^2 = a(ax + b) + b$$

Exercise 1 – If $xT^3 = x$, either $a^2 + a + 1 = 0$, or $(a - 1)x + b = 0$.

§ 8. **Multiplication** – We have already seen how to add two points of a plane, and how to multiply any point a by a point on the base-line, ρ. For the product of two points a and x we construct the triangle x, 0, y positively similar to the triangle 1, 0, a. Then (fig. 6)

$$y = ax = xa$$

This operation for a given a may be called a proportion, and y/x or a is a ratio.

The point a has the "polar co-ordinates" p and θ. We regard p as a positive stretch S about 0 which sends 1 into p, and θ as assigning a rotation R about 0, which sends p into a. These are commutative operations—that is, SR = RS. Applied to the point 1, SR gives the point a. Applied to any point x, SR gives xa.

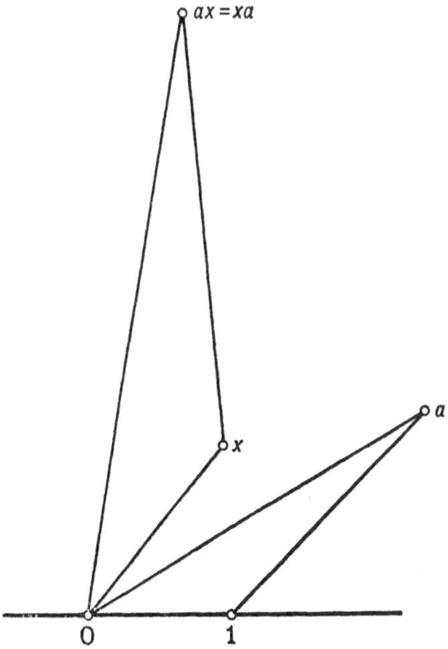

FIG. 6

Thus in terms of the polar co-ordinates p_1, θ_1 and p_2, θ_2 the product of two points has the polar co-ordinates p_1p_2 and $\theta_1+\theta_2$.

From this the commutative and associative laws for multiplication are true.

Since a parallelogram remains a parallelogram under S and R, the points 0, ba, $(b+c)a$, ca form a parallelogram. That is, $ba+ca=(b+c)a$.

All requirements are fulfilled, and a point (with reference to the points 0 and 1) is a number; or, if preferred, is the adequate representation of a number.

We call the circle with centre 0 on the point 1 the base-circle. On both the base-line and the base-circle are the points ± 1.

If we make the triangle 1, 0, ι positively similar to the triangle ι, 0, -1 then

$$\iota^2=-1$$

There are then two points ι, the points 1, ι, -1, $-\iota$ being successive vertices of a square. Selecting one of these two points as ι, we have assigned an arrow-head to the base-circle, and a positive sense of rotation.

The vertical line on 0, $\iota, -\iota$ is the vertical axis or axis of imaginaries. The number ι is rotation through a right angle in the positive sense. Any point on the vertical axis is by the law of proportion $\iota\eta$, where η is real. And by the parallelogram law any point of the plane is $\xi+\iota\eta$, where ξ and η are real. Here ξ and η are rectangular co-ordinates, abscissa and ordinate, of the point $\xi+\iota\eta$.

If now we denote a point on the base-ray by the positive number p and a point on the base-circle by t, then by the law of proportion the circle with centre 0 and radius p meets the ray from 0 to t at the point pt or tp. Thus, with the special numbers p and t, any point of the plane is named pt. This way of naming points is closely associated with polar co-ordinates. We may call it polar naming.

An essential property of a line in a plane is that it sets up a reflexion. That is, for any point x there is an image-point y such that the join is perpendicular to the line and is bisected by it. We may express this by saying that the plane can be folded on itself about the line; then x and y will coincide.

We denote the reflexion in the base-line by $y=\bar{x}$ (or $x=\bar{y}$). The points x and $\bar{x}$, images in the base-line, are said to be conjugate.

The base-line itself is then given by

$$x=\bar{x}$$

Thus a real number is one which is equal to its conjugate, or is self-conjugate.

By the law of proportion $x\bar{x}$ is a point on the base-ray. Such a number is positive. We call it the *power* of x (or of $\bar{x}$).

Again by the law of proportion, to find $1/x$ when x is given we make the triangle 1, 0, x positively similar to y, 0, 1. Then $xy=1$, or $y=1/x$. We call $1/x$ the reciprocal of x.

The conjugate of the reciprocal is then

$$y=1/\bar{x}$$

or

$$y\bar{x}=1$$

The two points x and y, where each is the conjugate of the reciprocal of the other, are then on a ray from 0, and at reciprocal distances from 0. Two such points are said to be inverse as to the base-circle or images in the base-circle.

When $x=y$ we have $x\bar{x}=1$, the equation of the base-circle. Thus the number t is the conjugate of its reciprocal. We call it (as a name meant abstractly) a *turn*. When we write a point x in the form pt, we call t the direction of x.

When $x-x_1=pt$, t is the direction of the line from x_1 to x. The undirected line on x and x_1 has two directions t and $-t$. The square t^2 is called the *clinant* of the line.*

The power of the two points, since $\bar{x}-\bar{x}_1=p/t$, is $p^2=(x-x_1)(\bar{x}-\bar{x}_1)$. The positive square root of this is the distance. It is written $|x-x_1|$ when convenient. The clinant t^2 is $(x-x_1)/(\bar{x}-\bar{x}_1)$.

§ 9. **Trigonometry** – Let us illustrate the use of direction as against angle. For a point on the base-circle, with the

* F. Franklin, *Am. Journal*, vol. 11.

number t and the angle θ, we call the abscissa $\cos\theta$ and the ordinate $\sin\theta$. Hence

$$t = \cos\theta + \iota\sin\theta$$

and

$$1/t = \cos\theta - \iota\sin\theta$$

So that $2\cos\theta = t + 1/t$, $2\iota\sin\theta = t - 1/t$.

Also

$$t^n = \cos n\theta + \iota\sin n\theta$$

$$2\cos n\theta = t^n + 1/t^n, \quad 2\iota\sin n\theta = t^n - 1/t^n$$

Again

$$t_1 t_2 = \cos(\theta_1 + \theta_2) + \iota\sin(\theta_1 + \theta_2)$$

and also

$$= (\cos\theta_1 + \iota\sin\theta_1)(\cos\theta_2 + \iota\sin\theta_2)$$

$$= \cos\theta_1\cos\theta_2 - \sin\theta_1\sin\theta_2 + \iota(\sin\theta_1\cos\theta_2 + \cos\theta_1\sin\theta_2)$$

Hence we have the addition theorems

$$\cos(\theta_1 + \theta_2) = \cos\theta_1\cos\theta_2 - \sin\theta_1\sin\theta_2$$

and

$$\sin(\theta_1 + \theta_2) = \sin\theta_1\cos\theta_2 + \cos\theta_1\sin\theta_2$$

Exercise 2 – Prove that

$$\sin(\beta - \gamma)\sin(\alpha - \delta) + \sin(\gamma - \alpha)\sin(\beta - \delta) + \sin(\alpha - \beta)\sin(\gamma - \delta) = 0.$$

Consider three points a_i. We have three edges which have the lengths or distances p_i, p_1 being $|a_2 - a_3|$. Let them have, as we go positively round the interior, the directions t_i, t_1 being that from a_2 to a_3. Then because we can add

$$p_1 t_1 + p_2 t_2 + p_3 t_3 = 0$$

and this carries with it the conjugate equation

$$p_1/t_1 + p_2/t_2 + p_3/t_3 = 0$$

Hence

$$p_1^2 = p_2^2 + p_3^2 + p_2 p_3 (t_2/t_3 + t_3/t_2)$$

Here $t_2/t_3 + t_3/t_2$ is twice the cosine of the exterior angle, $\pi - \alpha_1$, where α_1 is the interior angle.

Thus

$$p_1^2 = p_2^2 + p_3^2 - 2p_2 p_3 \cos\alpha_1$$

the law of cosines.

Also

$$p_1 = \kappa(t_2/t_3 - t_3/t_2)$$

where κ is symmetrical.

Thus

$$p_i = 2\iota\kappa \sin \alpha_i$$

the law of sines.

Exercise 3 – Determine κ in terms of p_i.

§ **10. Functions** – With algebra comes the forming of expressions containing constants and a variable. Thus we have

(1) the linear function, $y = \alpha_0 x + \alpha_1$
(2) the quadratic function, $y = \alpha_0 x^2 + 2\alpha_1 x + \alpha_2$
(3) the polynomial, $y = \alpha_0 x^n + n\alpha_1 x^{n-1} + \ . \ . \ . + \alpha_n$

and rational fractions such as

(4) $y = (\alpha_0 x + \alpha_1)/(\beta_0 x + \beta_1)$
(5) $y = (\alpha_0 x^2 + 2\alpha_1 x + \alpha_2)/(\beta_0 x^2 + 2\beta_1 x + \beta_2)$

and so on.

We may represent the number y on the same plane or on another plane. The plane is the same—that is, the base-point, base-ray, and base-circle are usually the same—unless the contrary is stated. In this case (1) is a *homology*, and (2) is a *homography*. These are fundamental because they have the group property. Namely, if $y = \alpha_0 x + \alpha_1$ and $z = \beta_0 y + \beta_1$, then $z = \gamma_0 x + \gamma_1$.

And so for (4) if y and x are one-to-one, and z and y are one-to-one, then z and x are one-to-one.

§ **11. The Derivative** – In the calculus of reals, the limit of the ratio of the change of y, the function, to the change of x is studied. The process is the same for the theory of functions, or calculus of numbers. Thus if $y = x^n$, and, for a given value x_1, y is y_1, then $(y - y_1)/(x - x_1)$ has the limit $nx_1{}^{n-1}$ where x tends to x_1. But this now means much more: x can tend to x_1 with any direction. The direction of $(y - y_1)/(x - x_1)$ at the limit is that of $nx_1{}^{n-1}$. If then two directions at x_1 make an angle θ, the two at y_1 make the same angle. This is the principle of isogonality. It breaks down when the derivative is 0 or ∞.

Thus, if the plane of x is divided into squares, the plane of y is divided into regions with right angles, by two systems of curves called orthogonal, or collectively a grid. Or, if the plane of x is divided by rays and concentric circles, we have two systems of curves in the plane of y which cut at right angles. These are a grid, provided the rays are taken at equal angles and the circles are properly spaced (fig. 7).

In fig. 7 we have two standard grids. The left-hand or chess-board one we call Euclidean. The other we call polar. The transition from the polar grid to the Euclidean is given in § 12; the reciprocal transition in § 13.

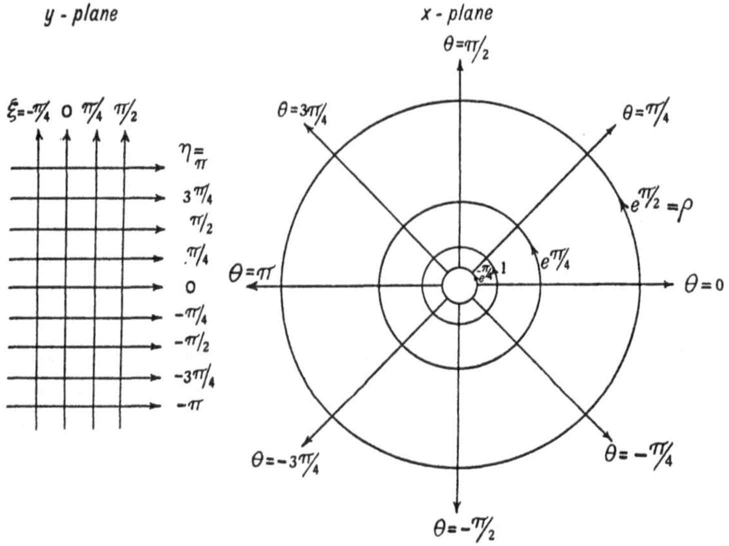

Fig. 7

The derivative being the same for numbers and for reals, we need not linger over the elementary processes for obtaining it for the above functions.

The technique of integration is the same for numbers and for reals. The consideration of the path of integration is here important; but in the cases here considered the paths are usually the base-line or the base-circle.

The above functions are all rational. In general the best way of extending the notion of a function is the power-series. Namely, in the neighbourhood of the point 0, y is given as

$$y = a_0 + a_1 x + a_2 x^2/2\,! + a_3 x^3/3\,! + \dots$$

the series being usually infinite.

Here a_0 is the value at $x=0$; a_1 is the first derivative at 0, a_0+a_1x is the first approximation; a_2 is the second derivative at $x=0$, $a_0+a_1x+a_2x^2/2\,!$ is the second approximation.

The series must converge when each term is replaced by its distance or absolute value.

Similarly, near a point a function may be

$$a_0+a_1(x-a)+a_2(x-a)^2/2\,!+\ .\ .\ .$$

where a_i is the ith derivative at a.

We note that the first approximation is a homology. Instead of the tangent line of a curve at a given point, for reals, the question is of the tangent homology of a function at a given point. Also when x and y are cogredient (that is, represented on the same plane) the direction at x and that at y meet, say at p, on a definite circle on both x and y. This circle is called the indicatrix. There are two special directions at x and two at y, for which the triangle x, z, y is isosceles.

§ **12. The Logarithm** - When we speak of the plane of numbers x, or simply the plane x, we suppose each individual x given either as $\xi+\iota\eta$ or as pt, according as we wish to add or to multiply.

If we name x by distance p and angle θ, we have for each point x many names; we avoided this by restricting θ to the principal angle. In fact we draw the ray 0 to -1—the negative ray—and refuse to cross it.

Consider now the square roots of x, $\pm\sqrt{x}$. The distance is $p^{\frac{1}{2}}$, the positive root of p. The angles are $\theta/2$ and $\theta/2+\pi$, where θ is the principal angle. We take as the principal square root that with angle $\theta/2$, and denote it by $n^{\frac{1}{2}}$ or $p^{\frac{1}{2}}t^{\frac{1}{2}}$. The region for $x^{\frac{1}{2}}$ is then half the plane—the right half. Strictly the ray 0, ι belongs to this half-plane, and the ray 0, $-\iota$ does not.

So we understand $x^{\frac{1}{4}}$, the principal fourth root of x. It lies in the sector $-\pi/4<\theta\leqslant\pi/4$. And so on.

But the limit of this process is simply the point 1.

We consider the limit of $n(x^{1/n}-1)$, in connection with the curve known as an equiangular spiral. This is the curve which arises from the point x, by marking all the points $.\ .\ .\ x^{-2}$, x^{-1}, 1, x, $x^2\ .\ .\ .$, these giving a fan of similar triangles; then the points . . . $x^{-\frac{3}{2}}$, $x^{-\frac{1}{2}}$, $x^{\frac{1}{2}}$, $x^{\frac{3}{2}}$ . . . The curve is then the limit of the fan; it represents all the points x^{ρ} where ρ is real. But we restrict ourselves to principal roots—that is, we consider the principal spire of the spiral as in fig. 8. When x is positive the curve is the positive ray; when x is a turn the curve is the base-circle. The curve in general is that continuous curve which cuts all rays from 0 at the same angle, say α, except when α is 0, in which case it is the base-ray itself. It is mechanically drawn by a rod on a ring at 0, carrying a toothed wheel whose axle is set at the angle $\pi/2 - \alpha$ to the rod.

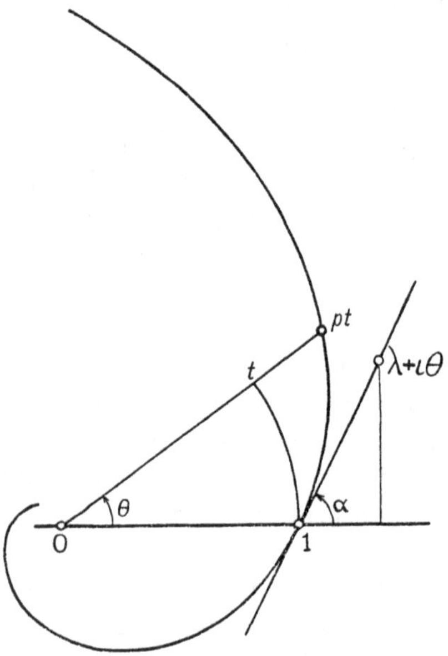

FIG. 8

Let now y be the limit (if there is one) of $n(x^{1/n} - 1)$. We know the direction of this limit, for it is that of the tangent to the curve at 1.

We regard y as in a superposed plane, with base-point 0 at $n = 1$.

Let now x be pt. Then

$$\begin{aligned} n(x^{1/n} - 1) &= n(p^{1/n}t^{1/n} - 1) \\ &= n(p^{1/n} - 1)t^{1/n} + n(t^{1/n} - 1) \end{aligned}$$

The limit of $n(t^{1/n} - 1)$ has the direction ι, and the distance θ, where θ is the principal angle of t. It is then $\iota\theta$.

Let the limit of $n(p^{1/n} - 1)$ be λ.

We have then $y = \lambda + \iota\theta$.

This point is known, for it is on the tangent line and has the ordinate θ. Its abscissa λ is then known. As x varies along the curve, y varies along the tangent. To the complete spire corresponds that part of the tangent for which $-\pi < \theta \leqslant \pi$.

Thus the whole plane of x becomes a strip of the plane of y.

We have $\lambda = \theta \cot \alpha$. It is manifestly proper to take the particular spiral with the angle $\pi/4$. This we call the logarithmic spiral. Then

$$\lambda = \theta$$

That is, the circle with centre 0 and radius p cuts this spiral at one point x, and the angle actually turned through in passing from 1 to x along the spiral is λ.

If the angle is 1, then p is the number e.

There is then associated with a number $x = pt$ the number $y = \lambda + \iota\theta$. We call y the logarithm of x, $y = \log x$; when θ is a principal angle, y is the principal logarithm.

In particular, when $\theta = 0$, $t = 1$, the number λ is $\log p$, the real logarithm.

Its value is shown by the logarithmic spiral, for it is then θ.

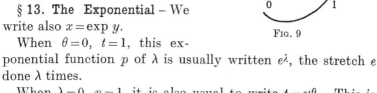

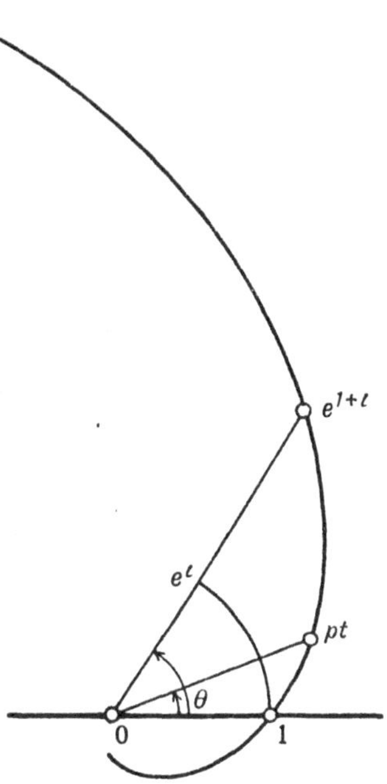

FIG. 9

§ 13. The Exponential – We write also $x = \exp y$.

When $\theta = 0$, $t = 1$, this exponential function p of λ is usually written e^{λ}, the stretch e done λ times.

When $\lambda = 0$, $p = 1$, it is also usual to write $t = e^{\iota\theta}$. This is e^{ι} done θ times. The number e^{ι} is rotation through unit angle. We may call this the standard rotation. So the number e is the standard stretch—that is, stretch with unit logarithm. What characterises this number e is the fact that $\lim n(e^{1/n} - 1)$ is 1. We may then write instead of $\exp y$, e^{y} or $e^{\lambda + \iota\theta}$; where $p = e^{\lambda}$, $t = e^{\iota\theta}$, $x = e^{\lambda + \iota\theta} = e^{\lambda}e^{\iota\theta} = e^{\iota\theta}e^{\lambda}$.

We may then write a number x in the form $pe^{\iota\theta}$; that is, in terms of its polar co-ordinates, distance and angle. This is a hybrid notation.

The function $e^{\iota\theta}$, the rotation, in terms of angle, has the period $2\iota\pi$. The function e^{λ} has then the same period. And so has $e^{\lambda+\iota\theta}$.

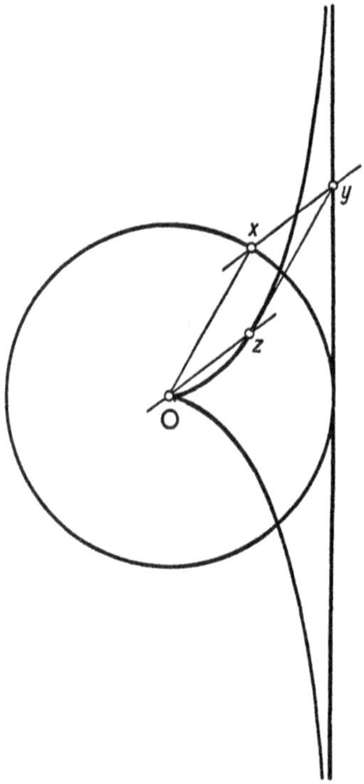

FIG. 10

In particular we have the remarkable relations of Euler:

$$e^{2\iota\pi}=1, \qquad e^{\iota\pi}=-1, \qquad e^{\iota\pi/2}=\iota$$

From the periodicity we see that any strip of the y-plane, between $\theta=\theta_0$ and $\theta=\theta_0+2\pi$, will give the whole x-plane.

If then the y-plane be rolled into a cylinder of radius 1, the strip becoming the surface of the cylinder, this cylinder maps into the plane of x.

If (anticipating § 23) we invert a sphere of radius 1 from its north pole on to its equatorial plane, then we have transferred the points x of the latter to points x of the sphere. The cylinder maps into this sphere.

We regard the sphere as inscribed in the cylinder touching it along the equator. If a point x of the sphere has longitude θ and south polar distance ϕ, then $p=\tan\phi/2$. The corresponding point y of the cylinder has the same longitude θ, and ordinate λ. Thus the relation is $\lambda=\log\tan\phi/2$.

This isogonal mapping of the sphere on to a cylinder or strip is Mercator mapping. The loxodromes or rhumb lines of the sphere become helices on the cylinder and lines on the plane.

The above relation between λ and ϕ is clearly shown in the instrument called a tractrix (fig. 10). We suppose a rod to move with one end on a line, the other end x moving in the direction of the rod. Analytically

$$\begin{aligned} x &= \lambda + t \\ dx &= d\lambda + dt \\ d\bar{x} &= d\lambda - dt/t^2 \end{aligned}$$

and we are to have

$$dx/d\bar{x} = t^2$$

Hence

$$\begin{aligned} d\lambda + dt &= t^2(d\lambda - dt/t^2) \\ d\lambda &= 2dt/(t^2 - 1) \\ &= dt/(t-1) - dt/(t+1) \\ \lambda &= \log(t-1) - \log(t+1) + \text{constant} \end{aligned}$$

Assigning $t = \iota$ to $\lambda = 0$,

$$\lambda = \log\left(\iota \frac{1-t}{1+t}\right)$$

Or, if $t = e^{\iota\phi}$,

$$\lambda = \log \tan \phi/2$$

CHAPTER III

THE EUCLIDEAN GROUP

§ 14. **Homologies** – Algebraically a homology H is the equation

$$y = ax + b$$

We consider here not the operations but the records—that is, x and y are thought of as points of the base-plane, in a correspondence H, and not as overlying points of overlying planes.

The homology H has a fixed point f given by

$$f = af + b$$

When $a = 1$ we say that the fixed point is at infinity, $f = \infty$. Taking the fixed point as 0, we have a canonical form for H, $y = ax$, except when a is 1. For this we may take $y = x + p$, where p is positive.

All homologies form a group. For if $z = cy + d$ and $y = ax + b$, then $z = cax + cb + d = ex + f$.

A homology is a stretch S when a is a real, a rotation R when a is a turn, a translation T when $a = 1$.

All translations form a group, since if $a = 1$ and $c = 1$, $ac = 1$. All stretches do not form a group, since if a is real and c is real, ac is real but may be 1. All stretches and all translations form a group.

So all rotations and all translations form a group—the group of planar displacements or of rigid motions.

§ 15. **Antilogies** – An antilogy A is the equation

$$\bar{y} = ax + \bar{b} \qquad (1)$$

It is the composition or product of H and a reflexion $\bar{y} = x$ in the base-line.

It has in general a fixed point f, given by

$$\bar{f} = af + \bar{b}$$

For then

$$f=\bar{a}\bar{f}+b$$

(2) $$f=a\bar{a}f+\bar{a}\bar{b}+b$$

With this as base-point, b is 0. Thus A has the form

$$\bar{y}=ax$$

Changing the unit of direction—that is, writing τx for x and $\tau^{-1}\bar{y}$ for $\bar{y}$—we have

$$\bar{y}=a\tau^2 x$$

We can choose τ^2 so that $a\tau^2$ is p. We thus get the canonical form of A,

(3) $$\bar{y}=px$$

This fails if $a\bar{a}=1$; that is, if A has the equation

$$\bar{y}=tx+\bar{b}$$

Changing the unit of direction this is

$$\bar{y}=x\pm(\lambda+\iota\mu)$$

and by a translation, $x=x'\pm\iota\mu/2$, we get

(4) $$\bar{y}=x+p$$

This is the canonical form for the general operation of an odd number of reflexions.

But if in (2) $\bar{a}=-b/\bar{b}$, then $a\bar{a}=1$ and f becomes arbitrary. Thus, for the antilogy

$$\bar{y}=-\bar{b}x/b+\bar{b}$$

or

(5) $$x/b+\bar{y}/\bar{b}=1$$

the conjugate equation is

$$\bar{x}/\bar{b}+y/b=1$$

In writing the conjugate we merely interchange x and y.

The points of the plane are then paired off by (5). It is the equation of a reflexion. When we make $x=y$, we get the equation of the line of fixed points,

(6) $$x/b+\bar{x}/\bar{b}=1$$

This is a self-conjugate equation. The point b is, from (5) when $y=0$, the image of the base-point in the line.

Antilogies are not a group, since the product of two is a homography. But they form a group—the Euclidean group.

Of this all the groups so far mentioned are subgroups. But there is also the important subgroup formed by all reflexions L, or antilogies of period 2. Two or any even number of reflexions give the positive or planar displacements, R or T. Three or any odd number give the negative displacements, for which the canonical form is

$$\bar{y} = x + p$$

We defer these to § 19, and consider next the composition or product of stretches, of rotations, and of general homologies.

§ 16. The Product of Stretches – The product or result of two stretches or of two rotations is naturally of importance. It is best viewed geometrically—that is, by a figure and not an equation.

A stretch S is (f, ρ), where f is the fixed point and ρ a real ratio. Let S_i be (f_i, ρ_i). We ask what are the relations on the fixed points f_i and the ratios ρ_i when $S_1S_2S_3 = I$.

The line $f_1 - f_2$ being fixed under S_1 and S_2 is fixed under S_3. The three fixed points are then on a line.

The stretch S_1 sends a circle C with centre y into a circle CS_1 with centre z. The stretch S_2 sends this latter into a circle with centre x. Thus this circle is CS_1S_2. The stretch S_3 sends this into the original circle, since $S_1S_2S_3$ is to be I. The product of ratios $\rho_1\rho_2\rho_3$ is then 1,

$$\rho_1\rho_2\rho_3 = 1 \tag{1}$$

But there is a further relation. We obtain it by placing y and therefore z at f_1. Then xS_3 is f_1 and f_1S_2 is x. We have then

$$f_1 - f_3 = \rho_3(x - f_3)$$

and

$$x - f_2 = \rho_2(x - f_2)$$

whence, eliminating x,

$$f_2 - f_3 + 1/\rho_3(f_3 - f_1) + \rho_2(f_1 - f_2) = 0 \tag{2}$$

This is one of three equivalent forms.

Exercise 1 – Deduce (2) algebraically from the equations :

$$\begin{aligned} S_1: &\ z-f_1=\rho_1(y-f_1)\\ S_2: &\ x-f_2=\rho_2(z-f_2)\\ S_3: &\ y-f_3=\rho_3(x-f_3) \end{aligned}$$

The proper thing is to introduce the auxiliary point f_0 on the fixed line, such that

$$\begin{aligned} f_2-f_0&=\rho_1(f_3-f_0)\\ f_3-f_0&=\rho_2(f_1-f_0)\\ f_1-f_0&=\rho_3(f_2-f_0) \end{aligned}$$

These equations satisfy both (1) and (2).

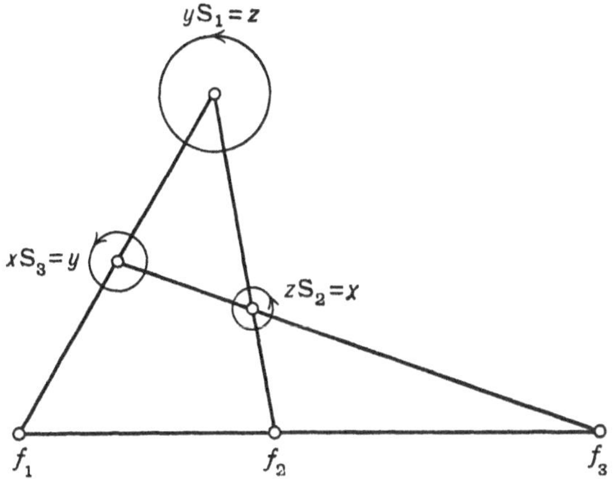

FIG. 11

This auxiliary point appears more naturally in the product of homologies (§ 18).

The figure for the product of two stretches is then a triangle and a fixed line, or, if preferred, the figure of three directed circles (fig. 11). In the former view the points of the plane are arranged in triangles with edges on the points f_i. One or three of the edges must be produced.

Exercise 2 – If we take two such triangles x, y, z and x_1, y_1, z_1 then $x-x_1$, $y-y_1$, $z-z_1$ are parallel. The corresponding edges $y-z$ and y_1-z_1 meet at f_1.

Consider the next case, $S_1S_2S_3S_4=I$. Here S_1S_2 and S_3S_4 are reciprocal stretches with the same fixed point f, the intersection of f_1-f_2 and f_3-f_4. Taking any point x, the points x, xS_1, xS_1S_2, $xS_1S_2S_3$ form a closed cycle. The diagonal $x - xS_1S_2$ is on f. So also the other diagonal $xS_1 - xS_1S_2S_3$ is on a fixed point f', the intersection of $f_2 - f_3$ and $f_4 - f_1$. We have then four lines, $f_1 - f_2$, $f_2 - f_3$, $f_3 - f_4$,

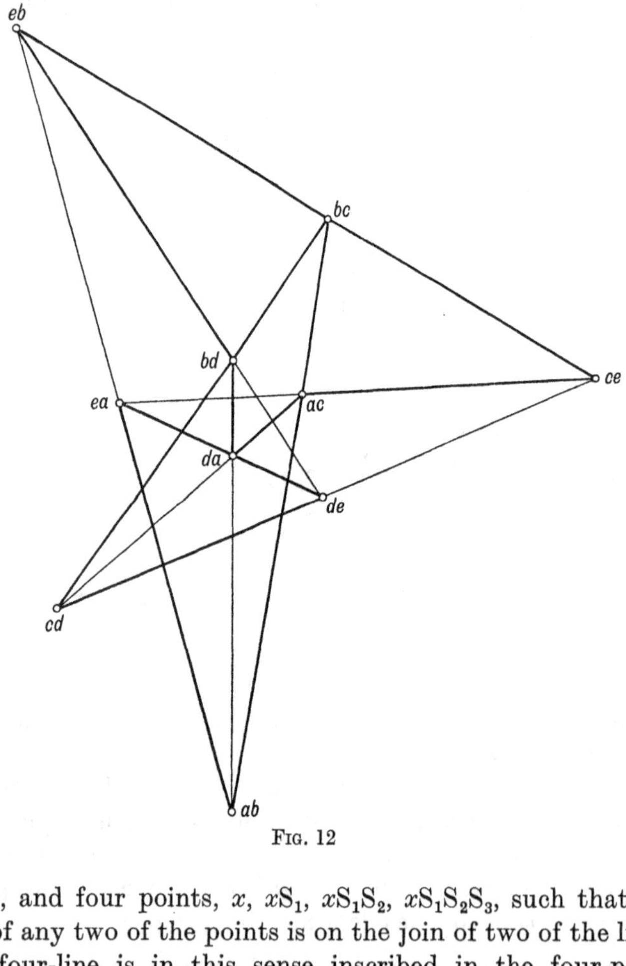

FIG. 12

$f_4 - f_1$, and four points, x, xS_1, xS_1S_2, $xS_1S_2S_3$, such that the join of any two of the points is on the join of two of the lines. The four-line is in this sense inscribed in the four-point. The complete figure is the Desargues configuration of ten points and ten lines, three points on each line and three lines on each point (fig. 12).

§ 17. **The Product of Rotations** – To compound rotations is simpler, for a rotation R is the product of two reflexions in lines L_1, L_2 which meet at, say, f, and one of these lines may be taken as any line on f.

If we denote R by (f, t) then t is the relative clinant of L_2 and L_1.

To compound R_1 and R_2 we take the line $f_1 - f_2$ or L_3 as the second line of R_1 and the first line of R_2. We have thus

$$R_1 = L_2L_3 \quad \text{and} \quad R_2 = L_3L_1$$

Then

$$R_1R_2 = L_2L_3{}^2L_1 = L_2L_1$$

Or symmetrically if

$$R_1 = L_2L_3, \quad R_2 = L_3L_1, \quad R_3 = L_1L_2$$

then

$$R_1R_2R_3 = I$$

Thus successive rotations about the vertices f_1, f_2, f_3 of a triangle, through twice the angles of the triangle, amount to

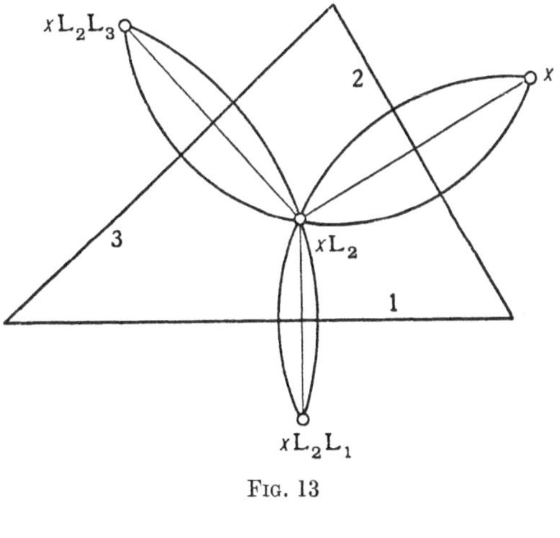

FIG. 13

identity. This is referred to as the triangle of rotations. Representing the rotations by arcs, we have a closed figure of three successive arcs which meet at a point (fig. 13).

We call this figure a trefoil. We may of course select arcs which intersect only when extended.

Exercise 3 – Draw a figure for the important special case when one rotation becomes a translation. The arcs are then similar.

If four rotations R_i amount to identity, then when represented by arcs ab, bc, cd, da, which intersect again successively at δ, a, β, γ (fig. 14), we have $R_1R_2 =$ arc $a\delta c$, $R_3R_4 =$ arc $c\beta a$, and therefore arc $a\delta c \times$ arc $c\beta a = I$. Hence $a\delta c\beta$ are on a circle. And conversely, if $a\delta c\beta$ are on a circle (in this order) then $R_1R_2R_3R_4 = I$. But then also $R_2R_3R_4 = R_1^{-1}$ and $R_2R_3R_4R_1 = I$. Therefore also $bad\gamma$ are on a circle.

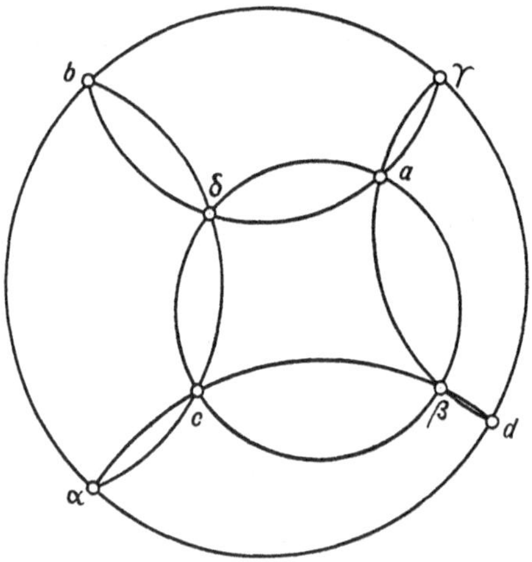

FIG. 14

If, then, we take four points $a\delta c\beta$ on a circle and draw any circles through $a\delta$, δc, $c\beta$, βa the remaining intersections of the successive circles are on a circle.

We have here a 4×2 scheme of eight points:

$$\begin{matrix} a & b & c & d \\ \alpha & \beta & \gamma & \delta \end{matrix}$$

such that any two of a row and the non-corresponding two of the other row are on a circle; this being not six, but only five conditions on the eight points.

Exercise 4 – A closed polygon of arcs represents a rotation R.

Let the arcs be a_1, a_2; a_2, a_3; . . . a_n, a_1.

Show that the angle of R is $-\Sigma\theta_i$ when θ_i is the exterior angle at the point a_i.

In particular, for a translation $\Sigma\theta_i = 0 \pmod{2\pi}$.

Exercise 5 – When all exterior angles are 0, the arc-polygon has continuity of direction. This is possible only for an even number of circles in successive contact. Show that for 4 circles having successive contact at a_i the four points a_i are on a circle.

Exercise 6 – If a_1, a_2, a_3 are on a line, and a_3, a_4, a_5 are on a line, and a_5, a_6, a_1 are on a line, the circles a_2, a_3, a_4; a_4, a_5, a_6; a_6, a_1, a_2 are on a point.

For extensions to higher spaces, see Study, *Math. Ann.*, vol. 36, and M. W. Haskell, *Grunert's Archiv*, 1903.

Exercise 7 – In general for $2n$ points a_i the successive arcs have continuity of direction when

$$(a_1 - a_2)(a_3 - a_4) \ldots (a_{2n-1} - a_{2n})/(a_2 - a_3)(a_4 - a_5) \ldots (a_{2n} - a_1)$$

is real. The first arc is then any arc joining a_1 to a_2. That is, the problem of drawing $2n$ arcs with continuity of direction is poristic; when it is possible, it is possible in an infinity of ways. (H. BATEMAN.)

§ 18. **The Product of Homologies** – Geometrically, a homology with centre C sends a circle into a circle, and a circle C on f into another circle CH on f. C and CH will meet again at d. Any point x of C is sent into a point xH of CH. The join of

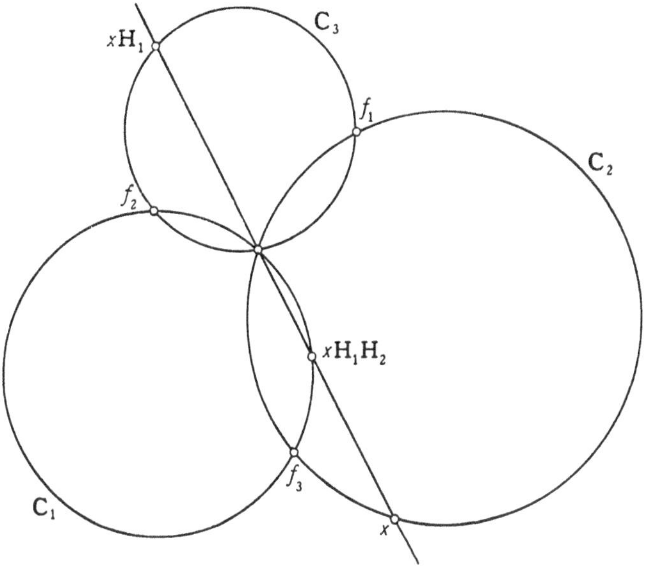

FIG. 15

these points will be on d, since the angles f, x, d and f, xH, d are given.

Thus the two circles, ordered, give a convenient picture of a homology.

Consider the three circles C_i meeting at d, and meeting again at f_1, f_2, f_3 (fig. 15). There are a homology H_1 sending C_2 into C_3, a homology H_2 sending C_3 into C_1, a homology H_3 sending C_1 into C_2. Thus the product of the three is identity, or

$$H_1H_2H_3 = I$$

The point x of C_2, when joined to d, gives the point xH_1 of C_3 and the point xH_1H_2 of C_1.

We show that three homologies whose product is I lead uniquely to the above figure. Consider three positively similar maps on a plane. There are three fixed points f_i, one for each two maps. And there is an auxiliary point a, such that for corresponding points of the maps

$$(x_2 - f_1)/(x_3 - f_1) = (f_3 - a)/(f_2 - a)$$
$$(x_3 - f_2)/(x_1 - f_2) = (f_1 - a)/(f_3 - a)$$
$$(x_1 - f_3)/(x_2 - f_3) = (f_2 - a)/(f_1 - a)$$

Take a as base-point. Then

$$x_1 f_1 + f_2 f_3 = x_2 f_2 + f_3 f_1 = x_3 f_3 + f_1 f_2$$

so that

$$\begin{aligned} x_1 &= f_2 + f_3 - f_2 f_3/u \\ x_2 &= f_3 + f_1 - f_3 f_1/u \qquad (1) \\ x_3 &= f_1 + f_2 - f_1 f_2/u \end{aligned}$$

When the variable u is f_1, then $x_2 = x_3 = f_1$, and

$$x_1 = f_2 + f_3 - f_2 f_3/f_1 = g_1, \text{ say,}$$

that is

$$f_1(f_1 - g_1) = (f_1 - f_2)(f_1 - f_3)$$

From (1),

$$x_1 - x_2 = (f_2 - f_1)(1 - f_3/u)$$
$$(x_1 - x_2)/(x_1 - x_3) = (f_2 - f_1)(u - f_3)/(f_3 - f_1)(u - f_2)$$

When x_1, x_2, x_3 are on a line, these are real.

Thence the angle f_2, u, f_3 is the angle f_2, f_1, f_3 (mod π), so that u is on the circle on f_1, f_2, f_3.

And

$$(x_1 - f_2)/(x_1 - f_3) = f_3(u - f_2)/f_2(u - f_3)$$

so that when a is on the circle f_i, the angle f_2, x_1, f_3 is a constant (mod π); that is, x_1 is on the circle f_2, g_1, f_3.

Thus, when in the three similar maps three corresponding points x_i are on a line, each point x_i is on a definite circle on two fixed points. The three circles meet at a point d (the director point) and all the lines are on d. For a purely geometrical account of this "theory of three similar figures" see Casey, *Sequel to Euclid*, 6th edition.

When the auxiliary point is the circumcentre of f_i, each homology becomes a rotation. Replacing f_i by a turn t_i, the equations (1) become

$$x_1 = t_2 + t_3 - t_2 t_3/u$$

and replacing u by its inverse y as to the base circle, we get the equations (1) in the secant-form

$$x_1 = t_2 + t_3 - t_2 t_3 \bar{y}$$

Exercise 8 – In this case the director point d is

$$d = t_1 + t_2 + t_3$$

The relation of the points a and d to the fixed points f_i is of interest.

The line f_1, g_1 has the equation

$$\frac{x - f_1}{f_1 - g_1} = \frac{\bar{x} - \bar{f}_1}{\bar{f}_1 - \bar{g}_1} = \text{a real}$$

Since

$$(f_1 - g_1)f_1 = (f_1 - f_2)(f_1 - f_3)$$
$$(x - f_1)f_1(f_2 - f_3) = \text{its conjugate} = \text{a real}$$

Hence for the point d on all the lines we have three equations such as

$$(d - f_1)f_1(f_2 - f_3) = \lambda_1$$

whence

$$\Sigma\lambda_1/(d - f_1) = 0$$

and also

$$\Sigma\lambda_1/f_1 = 0$$

The latter equation with its conjugate determines the real ratios $\lambda_1 : \lambda_2 : \lambda_3$. It follows that when f_i are the fixed points, a the auxiliary point, and d the director point, the latter two are connected mutually as roots of an equation

(2) $$\Sigma\lambda_i/(x - f_i) = 0$$

where the λ_i are real.

We meet this equation again in § 143.

§ **19. Negative Displacements** – An odd number of reflexions reduces to a rotation and one reflexion; hence to a reflexion L_1 and a translation; and finally, by resolving the translation across and along the line L_1, to a reflexion L and a translation T parallel to the line L. This is the canonical reduction for a negative displacement (§ 15). It may be thought of as a twist in a space along a screw whose axis is the line L, through two right angles. With this in mind we call the displacement LT a twist, and the absolute distance of T the pitch.

If we denote a twist (through π) by W, then $W = LT = TL$. We note that the mid-point of x and xW is on the axis L, and that W and W^{-1} have the same axis.

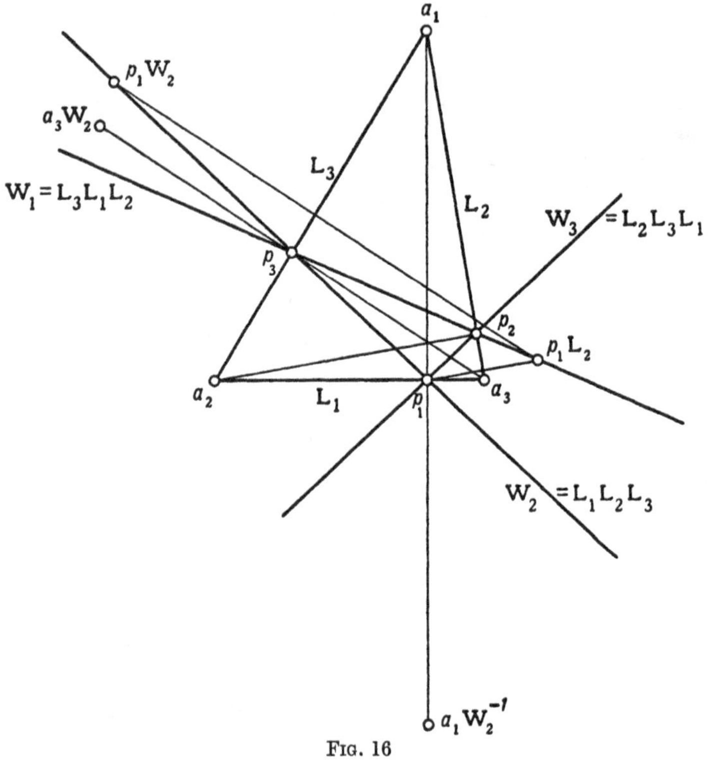

FIG. 16

The transform of an operation by a reflexion L_0 being the reflected operation, the transform of W,

$$L_0WL_0 = W'$$

is a twist whose axis is the image in L_0 of the axis of W.

Consider the product of reflexions in three lines forming a triangle a_1, a_2, a_3 (fig. 16). We have here

$$W_2 = L_1L_2L_3$$
$$W_2^{-1} = L_3L_2L_1$$

Call the vector from a vertex perpendicular to, and ending at, the opposite side an altitude. Let the end-points or feet of these altitudes be p_1, p_2, p_3. Since a_3 is on L_1 and L_2, a_3W_2

is a_3L_3, so that the mid-point of a_3 and a_3W_2 is p_3. This is a point on the axis of W_2. Since a_1 is on L_2 and L_3, $a_1W_2^{-1}$ is a_1L_1, so that the mid-point of a_1 and $a_1W_2^{-1}$ is p_1. This is a point on the axis of W_2^{-1}, that is of W_2. The axis of $L_1L_2L_3$ is then on p_3 and p_1. If we transform $L_1L_2L_3$ by L_1 we get $L_1L_1L_2L_3L_1$, that is $L_2L_3L_1 = W_3$. Its axis is on p_1 and p_2. Hence this line is the image of the line $p_3 - p_1$ in L_1. Similarly the transform of $L_2L_3L_1$ by L_2 is $L_3L_1L_2 = W_1$, with the axis $p_2 - p_3$.

To obtain the pitch of $L_1L_2L_3 = W_2$ we take p_1 and obtain $p_1L_1L_2L_3$, which is $p_1L_2L_3$ (fig. 16); p_1L_2 is on the line p_2, p_3, and $p_1L_2L_3$ is therefore on the line p_1, p_3. Hence for an acute-angled triangle the pitch, namely,

$$| p_1W - p_1 |$$

is the perimeter of the triangle p_1, p_2, p_3.

Exercise 9 – For a triangle with an obtuse angle at a_1 the pitch is

$$| p_1 - p_2 | - | p_2 - p_3 | + | p_3 - p_1 |$$

If now we regard an acute-angled triangle as an instrument, and overturn it first with the side 1 fixed, then with the side 2 fixed, and then with the side 3 fixed, we get fig. 17. The points p_3, p_1 of the first position (marked I) are in line with the points p_1, p_2 of the second position (marked L_1), and so on. The condition that the triangle be acute-angled is the condition that the axis of the twist shall cut the successive positions of the triangle. When this is fulfilled, the axis cuts the position I in p_1, p_3; reflecting the axis in sides 1 and 3 gives the triangle p_1, p_2, p_3.

This is a convenient planar picture of a twist, the reflexions being in the lines marked 11, 22, 33. The pitch is the perimeter of the triangle p_1, p_2, p_3. It is clear from this figure that the perimeter of p_1, p_2, p_3 is the minimum perimeter for inscribed triangles, for no other inscribed triangle would give a straight line.*

The two operations, of overturning the triangle on successive sides 1, 2, 3, and of reflecting the plane successively in lines 1, 2, 3, are reciprocal, and have the same axis.

* H. A. Schwarz, *Mathematische Abhandlungen*, vol. 2.

Let us construct fig. 18 for a pentagon analogous to fig. 17 for a triangle.

Regarding the pentagon as a card, we overturn it on sides 1, 2 . . . 5 successively, naming the positions I, L_1, . . . $L_1L_2L_3L_4L_5$. We write W for $L_1L_2L_3L_4L_5$. The mid-point

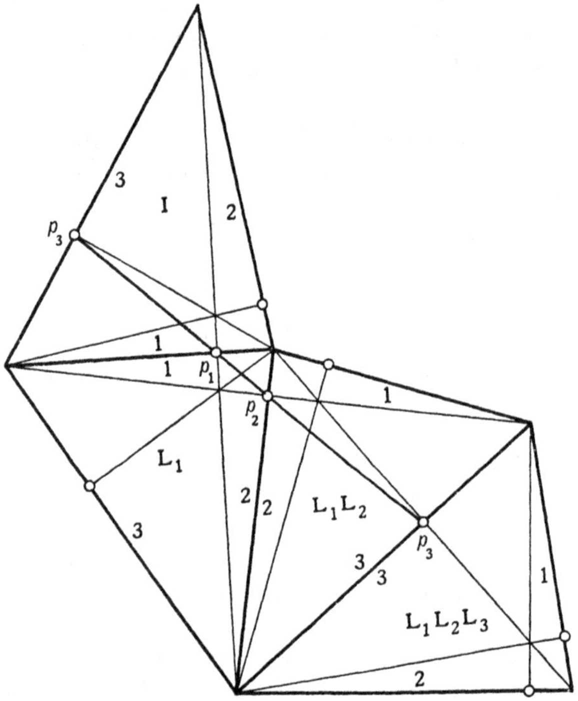

Fig. 17

of x and xW is on the axis of W. This axis is thus determined by the mid-points of say 23 and 23W, 45 and 45W.

The condition for an inscribed pentagon of minimum perimeter is that the axis of W shall cut all the successive positions of the given pentagon. Taking this case, the axis determines in position I the points p_1, p_5. Reflecting the axis in 1 and 5 of position I gives p_2 and p_4; and reflecting the line $p_1 - p_2$ in 2 or the line $p_5 - p_4$ in 4, gives p_3. We have then the pentagon p_1, p_2 . . . p_5 of minimum perimeter.

As before, the product of successive overturnings of the pentagon instrument, and the successive reflexions of the plane in the lines 1, 2, . . . 5, are reciprocal operations, with the same axis.

And so in general, for a polygon with an odd number of sides, say $2n+1$, the problem of constructing the minimum closed path which connects the successive sides is solved by

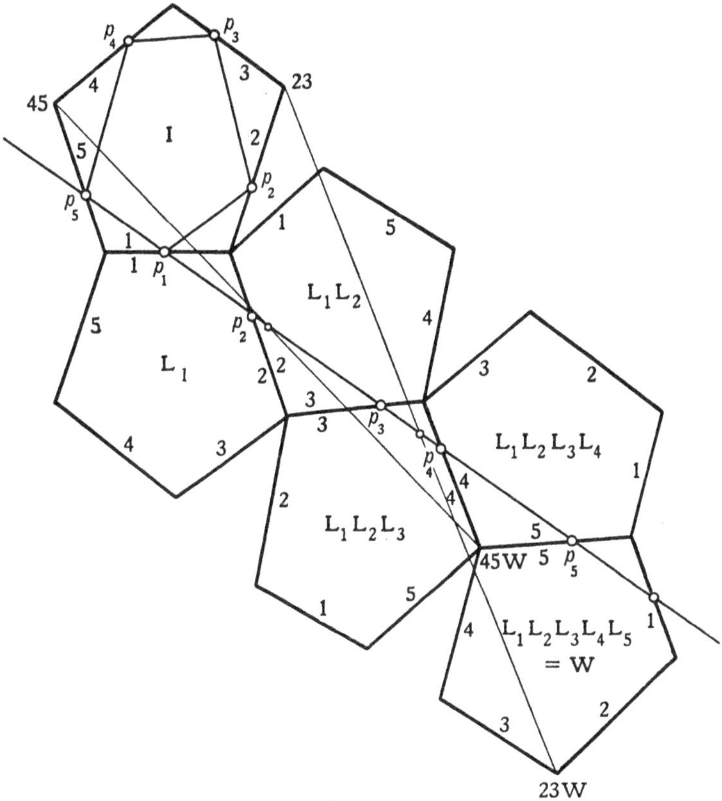

FIG. 18

constructing a figure analogous to fig. 18. The polygon P is displaced to PW, where $W = L_1L_2 \ldots L_{2n+1}$; the axis of W is determined by the mid-points of any convenient points x, xW and y, yW; and the points p_1, p_{2n+1} are cut by the axis on the sides 1, $2n+1$ of P; subject to the condition on P, that the axis of W shall cut all the successive positions of P.

Exercise 10 – The successive overturning or "rolling" of a concyclic quadrangle amounts to a translation.

CHAPTER IV

INVERSIONS

§ 20. **Cross-ratios** – There are two types of one-to-one correspondence of numbers, x and y, the homography

$$H \equiv \alpha xy + \beta x + \gamma y + \delta = 0$$

and the antigraphy

$$A \equiv \alpha x\bar{y} + \beta x + \gamma \bar{y} + \delta = 0$$

Each is determined by three values of x and the corresponding values of y. Thus two co-ordered sets of four points x_i and y_i are in a homography when

$$\begin{vmatrix} x_1y_1, & x_1, & y_1, & 1 \\ x_2y_2, & x_2, & y_2, & 1 \\ x_3y_3, & x_3, & y_3, & 1 \\ x_4y_4, & x_4, & y_4, & 1 \end{vmatrix} = 0$$

We write this determinant,

$$| \; x_iy_i, \quad x_i, \quad y_i, \quad 1 \; |$$

And the two sets of four points are in an antigraphy when

$$| \; x_i\bar{y}_i, \quad x_i, \quad \bar{y}_i, \quad 1 \; | = 0$$

There should then be a number connected with four points which is constant under H. All the homographies which send x_1 into y_1 and x_2 into y_2 are evidently given by

$$\frac{x - x_1}{x - x_2} = \kappa \frac{y - y_1}{y - y_2}$$

The homography which also sends x_3 into y_3 must have

$$\frac{x_3 - x_1}{x_3 - x_2} = \kappa \frac{y_3 - y_1}{y_3 - y_2}$$

Thus the homography which sends x_1, x_2, x_3 into y_1, y_2, y_3 is

$$\frac{(x - x_1)(x_2 - x_3)}{(x - x_2)(x_3 - x_1)} = \frac{(y - y_1)(y_2 - y_3)}{(y - y_2)(y_3 - y_1)}$$

If we write then for the four points x_i

$$\pi_1 = (x_2 - x_3)(x_1 - x_4)$$
$$\pi_2 = (x_3 - x_1)(x_2 - x_4)$$
$$\pi_3 = (x_1 - x_2)(x_3 - x_4)$$

where

$$\pi_1 + \pi_2 + \pi_3 = 0$$

the six ratios of the π_i are constant under H. These ratios with sign changed are the *cross-ratios*. The determinant $| x_i y_i, x_i, y_i, 1 |$ is then, to a numerical factor,

$$(x_2 - x_3)(x_1 - x_4)(y_3 - y_1)(y_2 - y_4) - (x_3 - x_1)(x_2 - x_4)(y_2 - y_3)(y_1 - y_4)$$

Exercise 1 – By considering the leading term $x_1 y_1 x_2 y_3$, show that the factor is -1.

Exercise 2 – The homography which sends $x = 0, \infty, 1$ into $y = 1, -1, \iota$ is

$$y = (1 + \iota x)/(1 - \iota x)$$

It sends the base line into the base circle.

Similarly for the determinant $| x_i \bar{y}_i, x_i, \bar{y}_i, 1 |$. Under an antigraphy a cross-ratio becomes its conjugate.

If r is a cross-ratio, then $1 - r$ and $1/r$ are cross-ratios. Hence the six cross-ratios, in terms of any one of them, r, are

$$\begin{array}{ccc} r & 1/(1-r) & (r-1)r \\ 1-r & r/(r-1) & 1/r \end{array}$$

Two of these are equal when

(1) $r = 0$, 1, or ∞, the scheme being $\begin{array}{ccc} 0 & 1 & \infty \\ 1 & 0 & \infty \end{array}$

(2) $r = \frac{1}{2}$, 2, or -2, the scheme being $\begin{array}{ccc} \frac{1}{2} & 2 & -1 \\ \frac{1}{2} & -1 & 2 \end{array}$

(3) $r^2 - r + 1 = 0$, the scheme being $\begin{array}{ccc} -\omega & -\omega & -\omega \\ -\omega^2 & -\omega^2 & -\omega^2 \end{array}$

In (1) one of the products π_i vanishes, and two of the four points coincide.

In (2) two of the products π_i are equal.

In (3) either

$$\pi_1 + \omega\pi_2 + \omega^2\pi_3 = 0$$

or

$$\pi_1 + \omega^2\pi_2 + \omega\pi_3 = 0$$

where $\omega = e^{2\iota\pi/3}$.

If we divide the four points into two pairs, say x_1, x_2 and x_3, x_4, then there are *two* cross-ratios,

$$r = (x_1 - x_3)(x_2 - x_4)/(x_1 - x_4)(x_2 - x_3) = -\pi_2/\pi_1$$

which we denote by $(x_1 x_2 \mid x_3 x_4)$, and $1/r$

Exercise 3 – $x = (x1 \mid 0\infty) = (1x \mid \infty 0) = (0\infty \mid x1) = (\infty 0 \mid 1x)$.

§ 21. **Inversions** – We consider in this chapter the anti-graphies of period 2—that is, such that $A^2 = I$. These are called inversions. Writing A as

$$(x - a)(\bar{y} - \bar{b}) = \kappa$$

we are to have also, always,

$$(y - a)(\bar{x} - \bar{b}) = \kappa$$

and therefore

$$(\bar{y} - \bar{a})(x - b) = \bar{\kappa}$$

Hence

$$a = b \quad \text{and} \quad \kappa = \bar{\kappa}$$

Thus an inversion is

$$(x - c)(\bar{y} - \bar{c}) = \text{real}$$

or, written with homogeneous coefficients,

$$\rho x\bar{y} - \bar{a}x - a\bar{y} + \sigma = 0$$

where ρ and σ are real.

A circle is given by an equation

$$(x - c)(\bar{x} - \bar{c}) = r^2$$

With this is given an inversion

$$(x - c)(\bar{y} - \bar{c}) = r^2$$

stating that the points x, y are on a ray from c, and that the product of their distances from c is the constant r^2.

Thus a circle sets up in the plane a mutual one-to-one correspondence of points, except perhaps when $x = c$, in which case $y = \infty$. We make the correspondence universal by choosing to regard ∞ as a point. We are thus making one exception to the axioms of algebra, for, in $ab = c$, c is no longer definite when $a = 0$ and $b = \infty$. Strictly 0 and ∞ are limits, not numbers; but we agree to regard them as numbers.

A line is now a circle on the point ∞, and inversion in the line is what we called reflexion. To include lines it is proper to replace the above equations by

$$\rho x\bar{x} - \bar{a}x - a\bar{x} + \sigma = 0$$

and

$$\rho x\bar{y} - \bar{a}x - a\bar{y} + \sigma = 0$$

where ρ and σ are reals. We have then a line when ρ is 0. The coefficients are homogeneous; only their ratios concern us. The expression or form has then the same generality as the equation; and we can denote an inversion simply by the form

$$\mathrm{C} \equiv \rho x\bar{y} - \bar{a}x - a\bar{y} + \sigma$$

But when in this we make $y=x$, to find the fixed points, we do not always get a circle. For

$$\rho\mathrm{C} = (\rho x - a)(\rho\bar{y} - \bar{a}) + \rho\sigma - a\bar{a}$$

and we have a circle only when the discriminant $\rho\sigma - a\bar{a}$ is negative.

We have then three types of inversions—hyperbolic, elliptic, parabolic, according as the discriminant is $-$, $+$, or 0.

Exercise 4 – Reflexions are of the hyperbolic type.

For the elliptic type we have

$$(x-c)(\bar{y}-\bar{c}) + r^2 = 0$$

There being no fixed points in the plane, we erect at c a normal to the plane and take on it the two points at distance r from c. These are the fixed points. We call them an elliptic pair or an extra pair.

For the parabolic or singular type we have

$$(x-c)(\bar{y}-\bar{c}) = 0$$

Here to $x=c$ corresponds an arbitrary y. Between the circle and the extra pair we have this intervening case; the circle has become a point taken twice or the extra pair have come together.

Thus with the general inversion is associated the self-conjugate form

$$\rho x\bar{x} - \bar{a}x - a\bar{x} + \sigma$$

This defines a curve—a bilinear curve—which is a general word for the three types—circle, extra pair, double point.

In general we associate with any self-conjugate form in x and $\bar{x}$ the word curve. If we replace $\bar{x}$ by $\bar{y}$ (or x by y) we get the transformation associated with the curve—a generalisation of inversion.

§ 22. **Inversors** - An appropriate instrument for inversion is a linkwork or assemblage of jointed rods. The simplest forms are the Peaucellier cell (fig. 19 (a)), and Hart's axial

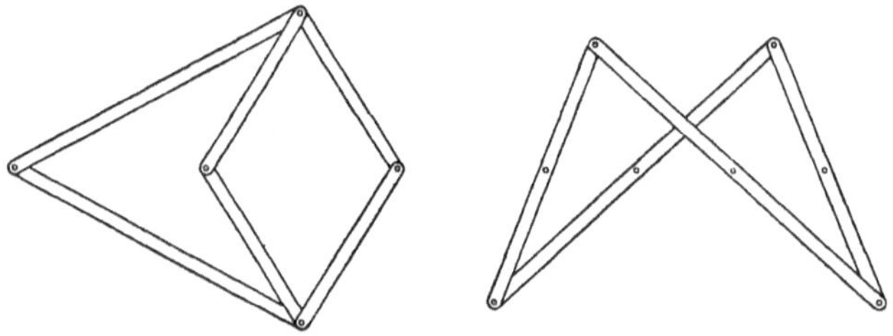

(a) FIG. 19 (b)

quadrangle or contra-parallelogram (fig. 19 (b)). In the former we have a rhombus x, p, y, q and a point c on the diagonal line $x-y$.

Then

$$|c-p|^2-|x-p|^2=(c-x)(\bar{c}-\bar{y})$$

Thus the rhombus of jointed rods, with the two bars $c-p$ and $c-q$ to the fixed point c, is an inversor, x and y being inverse points.

Hart's apparatus contains four bars only. It is a quadrangle a, b, $\bar{a}$, $\bar{b}$, where the axis of a and $\bar{a}$ is the axis of b and $\bar{b}$. This common axis is taken here as the base-line.

Let α, β, γ, δ be the mid-points of the bars. Then

$$\alpha-\beta=\delta-\gamma=(a-\bar{a})/2$$

and

$$\alpha-\delta=\beta-\gamma=(b-\bar{b})/2$$

By Ptolemy's theorem for four points on a circle,

$$|a-\bar{a}|\,|b-\bar{b}|+|a-b|\,|\bar{a}-\bar{b}|=|a-\bar{b}|\,|\bar{a}-b|$$

Hence, for all shapes of the given four-bar,

$$|\alpha-\beta|\,|\alpha-\delta|=\tfrac{1}{4}(\mu^2-\lambda^2)$$

and

$$|\beta-\alpha||\beta-\gamma|=\tfrac{1}{4}(\lambda^2-\mu^2)$$

where the lengths of the bars are λ, μ, λ, μ.

Thus, if we fix one of the points α, β, γ, δ, the two adjacent points will be in an inversion.

Exercise 5 – In Hart's inversor, if four points marked one on each bar be once in a line they will always be in a line.

For an account of inversors see A. B. Kempe, *How to Draw a Straight Line* (London, Macmillan, 1877).

§ 23. **Properties of an Inversion** – If we denote the distance from c to x by δ_{01} and from c to y by δ_{02}, then, for an inversion,

$$\delta_{01}\delta_{02}=\kappa$$

For a second inversion with the same centre,

$$\delta_{02}\delta_{03}=\kappa_1$$

Hence

$$\delta_{01}/\delta_{03}=\kappa/\kappa_1$$

This is a stretch. Thus a stretch is the product of two concentric inversions.

Conversely a stretch and a concentric inversion are commutative operations whose product is an inversion with the same centre.

Euclid's theorem that the product of segments on a secant from c to a circle C is a constant shows that a circle can be inverted into itself with any point not on it as centre. If c be outside, the inversion is hyperbolic and the circle of fixed points is orthogonal to the circle C.

If c be inside, the inversion is elliptic. The two fixed points are extra points on the sphere of which C is a great circle.

Since an inversion with centre c sends the circle C into itself, and a stretch with centre c sends C into another circle, any inversion whose centre is in the plane of a circle sends that circle into a circle. Lines here are included as circles on the point ∞.

Exercise 6 – Under $x\bar{y}=1$, $\rho x\bar{x}-\bar{a}x-a\bar{y}+\sigma$ becomes $\rho-\bar{a}y-a\bar{y}+\sigma y\bar{y}$.

But the same is true for any point c. For inversions and stretches apply equally to a Euclidean space. Thus an inversion sends a sphere into a sphere (planes being included as spheres on the point ∞). Hence an inversion in a space sends a circle into a circle, for it sends the intersection of two spheres into the intersection of two spheres.

An inversion with centre c sends two points x and x_1 into two points y and y_1, the four points being on a circle. If x_1 tends to x along this circle, then y_1 tends to y. At the limit the angle cxx_1 is the supplement of the angle cyy_1. Hence to two directions at x_1 at an angle θ correspond two directions at y at an angle $-\theta$. Thus angles are negatively equal, or reversed under an inversion.

In a space, angles are unaltered in size. For on the base x, y we have two isosceles triangles, say x, y, z and x, y, z'. The angles z, x, z' and z, y, z' are then equal.

Since

$$(x_1 - \rho)(\bar{y}_1 - c) = \kappa$$

and

$$(x_2 - c)(\bar{y}_2 - c) = \kappa$$

$$\bar{y}_1 - \bar{y}_2 = \kappa(x_2 - x_1)/(x_1 - c)(x_2 - c)$$

Thus the distance δ_{12} from x_1 to x_2 becomes δ_{12}' where

(1) $$\delta_{12}' = -\kappa\delta_{12}/\delta_{01}\delta_{02}$$

For instance, the relation

$$\delta_{12}' + \delta_{23}' = \delta_{13}'$$

which connects three points of a line, becomes under (1)

$$\delta_{12}\delta_{34} + \delta_{20}\delta_{14} = \delta_{13}\delta_{24}$$

which is Ptolemy's theorem for four points on a circle.

Exercise 7 - Three points are sent into the vertices of an equilateral triangle by an inversion whose centre is given by

$$\delta_{01}\delta_{23} = \delta_{02}\delta_{31} = \delta_{03}\delta_{12}$$

The effect of an inversion is to change a cross-ratio into its conjugate.

Thus to say that four points are on a circle is to say that they can be inverted each into itself. In this case, then, all cross-ratios are real. The four points are now ordered. By an inversion they are on a line. For four points x_i on a line in the order $x_1x_2x_3x_4$, π_1 is positive, π_2 is negative, π_3 is positive. Thus of the cross-ratios four are negative and two positive. These latter are the cross-ratios of the alternate pairs x_1, x_3 and x_2, x_4.

For four points in general an even number of inversions leaves a cross-ratio absolutely unaltered.

§ 24. **Normal Circles** – Two inversions C_1 and C_2 are in general not commutative. The product C_1C_2, where

$$C_1 \equiv \rho_1 x\bar{y} - \bar{a}_1 x - a_1\bar{y} + \sigma_1$$
$$C_2 \equiv \rho_2 z\bar{y} - \bar{a}_2 z - a_2\bar{y} + \sigma_2$$

is

$$(1) \qquad \begin{vmatrix} \rho_1 x - a_1 & \bar{a}_1 x - \sigma_1 \\ \rho_2 z - a_2 & \bar{a}_2 z - \sigma_2 \end{vmatrix} = 0$$

and this is symmetrical in x and z only when the coefficients of x and z are the same—that is,

$$p_{12} \equiv \rho_1\sigma_2 - \bar{a}_1 a_2 - a_1\bar{a}_2 + \sigma_1\rho_2 = 0$$

This expression is the bilinear invariant or power of C_1 and C_2.

Consider two circles C_1 and C_2, for which the power vanishes. For a point x on C_1, $xC_1 = x$. Hence $xC_1C_2 = xC_2$. This is to be xC_2C_1. Hence xC_2 is also on the circle C_1. The circle C_1 is then sent into itself by inversion in C_2; that is, the two circles are orthogonal or normal.

The equation (1) gives for two circles two points for which

$$x = z = f \text{ say}$$

For these,

$$fC_1C_2 = fC_2C_1$$

When C_1 and C_2 intersect, these two points f are the intersections. For then $fC_1 = f$ and $fC_2 = f$, so that $fC_1C_2 = f$.

When C_1 and C_2 do not intersect, the two points f are images in each circle. That is, $f_1C_1 = f_2$ and $f_2C_2 = f_1$, so that $f_1C_1C_2 = f_1$.

Thus the points common to two circles are the intersection when they intersect, and the common image-pair when they do not.

Taking one of the two points as ∞, we have the canonical form of two circles—two lines or two concentric circles.

Exercise 8 – The power of two lines $\bar{a}_i x + a_i \bar{x} = \sigma_i$ is $-(a_1\bar{a}_2 + a_2 a_1)$. This vanishes when the lines are normal or perpendicular.

Exercise 9 – If two circles intersect at 0, then $\sigma_1 = \sigma_2 = 0$. The power is $-(a_1\bar{a}_2 + a_2\bar{a}_1)$. The tangents at 0 are $\bar{a}_2 x + \bar{a}_1 \bar{x} = 0$. Hence the tangents are at right angles.

Exercise 10 – Solve the equations $x\bar{y} = 1$, $x + \bar{y} = 2\mu$.

§ **25. The Lune and the Ring** – Two circles, or generally two inversions, are then said to be orthogonal or normal when the power vanishes.

If the two become the same, the power is twice the discriminant,

$$p_{11} = 2(\rho_1\sigma_1 - a_1\bar{a}_1)$$

If we divide p_{12} by both square roots, $\sqrt{p_{11}}\ \sqrt{p_{22}}$, we get the fundamental constant κ_{12} (under inversions) for two circles. In terms of the radii r_1, r_2, and the distance δ_{12} of the centres, we have for C_1 with centre 0

$$x\bar{x} = r_1^2$$

and for C_2 with centre δ_{12}

$$(x - \delta_{12})(\bar{x} - \delta_{12}) = r_2^2$$

whence

$$p_{12} = \delta_{12}^2 - r_1^2 - r_2^2$$
$$p_{11} = -2r_1^2$$
$$p_{22} = -2r_2^2$$

whence

$$\kappa_{12} = \pm(\delta_{12}^2 - r_1^2 - r_2^2)/2r_1r_2$$

Exercise 11 – Two concentric circles with positive radii r_1, r_2 can be inverted into two equal circles with radius r. Prove that the distance of the centres is given by

$$d/r = \sqrt{r_1/r_2} + \sqrt{r_2/r_1}$$

A radius is positive or negative according as we give an arrow-head to a circle in the sense of ι or of $-\iota$.

Consider two circles which intersect. Let both be described positively. If C_2 makes with C_1 an angle θ (mod 2π) at one intersection, it makes the angle $-\theta$ at the other. Under an inversion the angles become $-\theta$ and θ. The function $\cos\theta$ is indicated, for $\cos\theta = \cos(-\theta)$. We have from the law of cosines

$$\cos\theta = (\delta_{12}^2 - r_1^2 - r_2^2)/2r_1r_2$$

Under an inversion r_1 or r_2, or both, may become negative, but the formula shows that when r_1r_2 changes sign, then also $\delta_{12}^2 - r_1^2 - r_2^2$ does. We may take then for intersecting directed circles,

$$\kappa_{12} = \cos \theta$$

The circles divide the plane into regions called *lunes*, or rather into two pairs of complementary lunes.

When two circles do not intersect, they include a ring. If the circles are external, the ring includes the point ∞. We consider all the arcs orthogonal to both. These terminate at the image pair f_1, f_2. Any arc meets C_1 say at x_1 and C_2 say at x_2. Consider the cross-ratio

$$\frac{(x_1 - f_1)(x_2 - f_2)}{(x_1 - f_2)(x_2 - f_1)}$$

First, it is real because the points are on a circle. Second, it is the same for all arcs. For inversion in any of the arcs will not alter it, but will send an arc $f_1x_1x_2f_2$ into another, say, $f_1y_1y_2f_2$. Third, it is positive, for it is manifestly positive for the segment $f_1 - f_2$.

Of this positive number, which is the same for all arcs, we take the logarithm, λ. This is the hyperbolic distance from C_1 to C_2, the same along all arcs.

Let us take the canonical case where the circles are concentric. Here it is proper to take the circles as oppositely described, so that if r_2 be positive, r_1 is negative.

We have $f_2 = \infty$ and take $f_1 = 0$. Then

$$e^{-\lambda} = x_2/x_1 = -r_2/r_1$$
$$e^{\lambda} = x_1/x_2 = -r_1/r_2$$

$$\tfrac{1}{2}(e^{\lambda} + e^{-\lambda}) = -\frac{r_1^2 + r_2^2}{2r_1r_2}$$

and this is the value of κ_{12} with $\delta_{12} = 0$.

Hence, in general, the relation between the logarithm of the ring, λ, and the constant κ_{12} is

$$\kappa_{12} = \tfrac{1}{2}(e^{\lambda} + e^{-\lambda})$$

This is the hyperbolic cosine of λ, $\cosh \lambda$.

Thus for the ring we have cosh λ, whereas for a lune we had cos θ.

For the angle one makes instinctively a numerical estimate. It is useful, though not so easy, to do so for the logarithm.

One way is to take a semicircle on f_1, f_2. To the tangents to C_1 at x_1 and to C_2 at x_2 we apply a standard tractrix with base parallel to $f_2 - f_1$. We thus get a distance along the base of the tractrix, which is λ.

The ring may also be measured inversively by an angle (or the lune by a logarithm).

Taking the canonical ring formed by concentric circles, of radii 1 and p, we inscribe a circle in the ring. It subtends at the centre an angle θ.

The tangent to it from the common centre is $p^{1/2}$.

Hence

$$\sec \theta/2 = \tfrac{1}{2}(p^{1/2} + p^{-1/2})$$

or since $p = e^{\lambda}$,

$$\sec \theta/2 = \cosh \lambda/2 \tag{1}$$

This, then, is the condition that a lune of angle θ and a ring

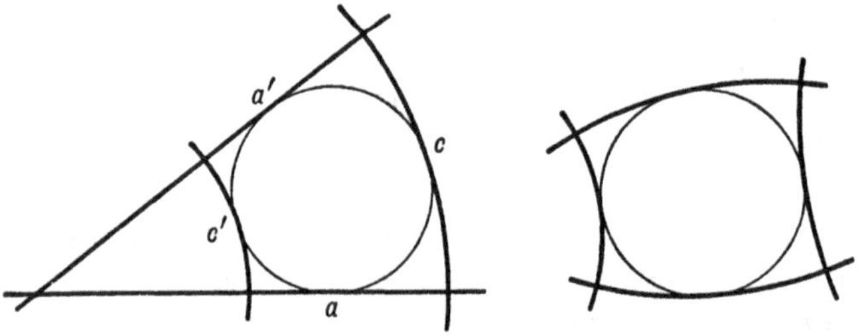

FIG. 20

of logarithm λ, the two being orthogonal, have an inscribed circle (fig. 20).

They form a polar square. By successive inversions in the two elliptic sides (which belong to the ring) we get fig. 21 (a). By successive inversions in the hyperbolic sides (which belong to the lune) we get fig. 21 (c). There is here closure when θ is commensurable with π. The intervening case (fig. 21 (b)) is that of a strip where the given circles touch. The figures illustrate the three fundamental measurements—the hyperbolic, the elliptic, and the parabolic.

Exercise 12 – If $\mu = {}^{\lambda/2}$ and $t=e^{\iota\theta/2}$, then (1) is

$$(\mu+1/\mu)(t+1/t)=4$$

Show that the join of the real, μ, and the turn, t, is on one or other of the points $\pm\iota$.

Exercise 13 – Writing $\mu t = x$, the equation is

$$x+x+1/\bar{x}+1/\bar{x}=4$$

Draw the curve. It is the inverse of a rectangular hyperbola as to an apse.

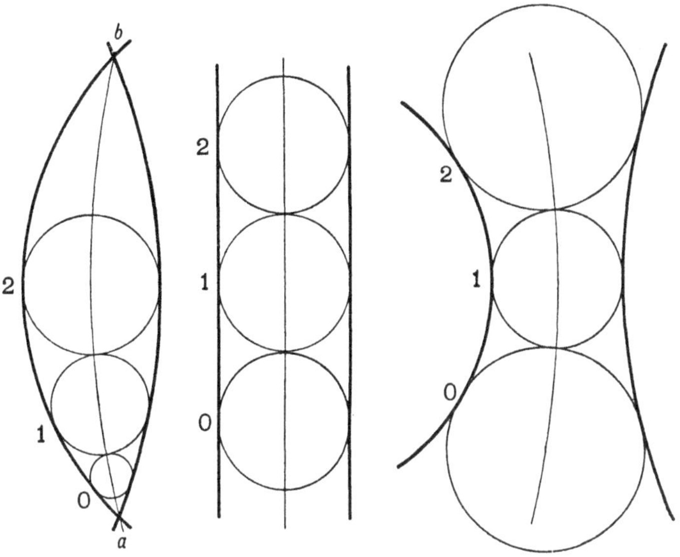

(a) lune (b) strip (c) ring

FIG. 21

§ 26. **The Canonical Form** – When four points x_i are a circle C_1 there are three inversions which interchange them. For we can take as centre the intersection of, say, the secants x_1-x_2 and x_3-x_4. Two of these inversions are hyperbolic, and thus have circles of fixed points normal to C_1.

Let then C_1 be the circle, in which x_2, x_3 and x_1, x_4 are images; and C_3 be the circle in which x_1, x_2 and x_3, x_4 are images.

Then C_1C_3 sends x_1 into x_3, x_3 into x_1, x_2 into x_4, and x_4 into x_2. And C_3C_1 does the same.

Hence the circles C_1, C_3 are normal. They meet at two

points, which we now take to be 0 and ∞ (by an auxiliary inversion). The circles are now normal lines, and the points are obtained by reflexion of one of them in these lines. The four-point has become a rectangle. This is a canonical form for four points on a circle.

In this canonical form we take the inversion forms as

$$\begin{aligned} C_1 &: \iota(x-\bar{y}) \\ C_2 &: x\bar{y}-1 \\ C_3 &: x+\bar{y} \\ C_0 &: \iota(x\bar{y}+1) \end{aligned}$$

where

$$C_0^2+C_1^2+C_2^2+C_3^2=0$$

Here C_0 is the elliptic form, and its fixed points are an extra pair. We may call them the north and south poles on the base-sphere, of which the base-circle is the equator.

With the four forms C_i we can express any inversion form as

$$\lambda_0 C_0+\lambda_1 C_1+\lambda_2 C_2+\lambda_3 C_3$$

where the λ_i are real.

The relations between the Cartesian co-ordinates of a point in a plane and those of a point on a sphere may be written

$$\begin{aligned} \xi_1 &= \xi^2+\eta^2-1 \\ \xi_2 &= 2\xi \\ \xi_3 &= 2\eta \\ \xi_0 &= \xi^2+\eta^2+1 \end{aligned}$$

So that

$$\xi_1^2+\xi_2^2+\xi_3^2=\xi_0^2$$

Here ξ_i/ξ_0 are Cartesian co-ordinates in a flat or Euclidean space.

A curve in the plane becomes a section of the sphere by a surface

$$f_n(\xi_0\xi_1\xi_2\xi_3)=0$$

But we are concerned only with the section, so that f may be modified by

$$(\xi_1^2+\xi_2^2+\xi_3^2-\xi_0^2)f_{n-2}(\xi_0\xi_1\xi_2\xi_3)$$

Thus the number of terms in f_n is

$$(n+1)(n+2)(n+3)/6-(n-1)n(n+1)/6$$

that is,

$$(n+1)^2$$

In the same way, if we invert a flat space from a point outside it, we may write

$$\begin{aligned}\xi_1 &= \xi^2+\eta^2+\zeta^2-1\\ \xi_2 &= 2\xi\\ \xi_3 &= 2\eta\\ \xi_4 &= 2\zeta\\ \xi_0 &= \xi^2+\eta^2+\zeta^2+1\end{aligned}$$

so that

$$\xi_1^2+\xi_2^2+\xi_3^2+\xi_4^2=\xi_0^2$$

Here ξ_i/ξ_0 are Cartesian co-ordinates in a flat four-way. A surface in the space becomes a section of the hyper-sphere by $f_n(\xi_0\xi_1\xi_2\xi_3\xi_4)$. But the number of terms of the section is

$$(n+1)(n+2)(n+3)(n+4)/24-(n-1)n(n+1)(n+2)/24$$

that is,

$$(n+1)(n+2)(3n+2)/6$$

This, then, is the number of terms in an algebraic surface in a flat space, when inversively considered, in terms of its inversive degree n. For the sphere $n=1$, and the number is 5. That is, a sphere can be expressed in terms of 5 spheres, for instance, in terms of $\xi^2+\eta^2+\zeta^2\pm 1$, ξ, η, ζ. For a cyclide $n=2$, and the number is 14.

Between six spheres, or between 15 cyclides, there must be a linear identity.

CHAPTER V

QUADRATICS

§ 27. **The Bilinear Invariant** – Given two points a_i of a plane, we have all the circles on the points and all the circles about the points (that is, as to which a_i are inverse points). Inversions in the former circles leave a_1 and a_2 individually fixed; inversions in the latter interchange a_1 and a_2.

Through a point x passes a circle of each set, and the two meet again at y. Thus the points of the plane are paired off.

The cross-ratio $(xy \mid a_1a_2)$ is a real because the two pairs are concyclic, and it is a turn because the pairs are anticyclic. It is then -1; if it were 1, two points would coincide. We have then, as the equation of the pairing,

$$(x-a_1)(y-a_2)+(x-a_2)(y-a_1)=0$$

or

$$2xy-(a_1+a_2)(x+y)+2a_1a_2=0$$

or, say,

$$\alpha_0 xy+\alpha_1(x+y)+\alpha_2=0$$

The pairing is a homography of period 2. This is called a quadratic involution or polarity. The fixed points are a_1 and a_2. They are given by the quadratic form

$$\alpha_0 x^2+2\alpha_1 x+\alpha_2$$

The relation of the pair x, y and the pair a_i is a mutual one. The two pairs are said to be normal, or apolar, or harmonic. So the two quadratic forms which give them are said to be normal, or apolar, or harmonic. If x and y are zeros of $\beta_0x^2+2\beta_1x+\beta_2$, the condition is

$$\alpha_0\beta_2-2\alpha_1\beta_1+\alpha_2\beta_0=0$$

This expression is the bilinear invariant or, as a typical invariant, the *power* of the two quadratics or of the two pairs of points x_1, x_1' and x_2, x_2'. When it vanishes the two quadratics or the two pairs are said to be harmonic or apolar or normal. We call them normal.

All pairs normal to a given pair are in an involution. For x and y are given by

$$z^2 - z(x+y) + xy$$

and this quadratic is normal to $\alpha x^2 + 2\beta x + \gamma$ if

$$\alpha xy + \beta(x+y) + \gamma = 0$$

Let the pairs of points be x_1, x_1' and x_2, x_2'.

Since

$$\alpha_i(x_i + x_i') = -2\beta_i, \qquad \alpha_i x_i x_i' = \gamma_i$$

the power is to an immaterial factor,

$$p_{12} \equiv (x_1 + x_1')(x_2 + x_2') - 2x_1x_1' - 2x_2x_2'$$

When this vanishes the pairs are normal.

Suppose that $x_2' = \infty$. Then $x_1 + x_1' = 2x_2$; that is, x_2, the polar of ∞ as to x_1, x_1', is the mid-point of x_1 and x_1'.

Again let $x_2 = 1$, $x_2' = -1$, $x_1 = \iota$. Then $x_1' = -\iota$. The two normal pairs are the diagonal points of a square. This is a convenient canonical form for normal pairs.

Exercise 1 – For normal pairs with $x_1 = 0$,

$$2/x_1' = 1/x_2 + 1/x_2'$$

and therefore, in general,

$$\frac{2}{x_1 - x_1'} = \frac{1}{x_1 - x_2} + \frac{1}{x_1 - x_2'}$$

Exercise 2 – Geometrically, a circle C pairs off the points of a space in which it lies. There is on any point x a sphere S containing C, and a pencil of spheres orthogonal to C, giving a circle. This circle meets the sphere S again at y. But analytically x and y are not numbers but quaternions.

Geometrically (that is, inversively) normal pairs are first on a circle Ω; and, second, the arc x_1, x_1' normal to Ω and the arc x_2, x_2' normal to Ω are themselves normal. There is then an inversion with x_1, x_1' fixed which interchanges x_2 and x_2', and an inversion with x_2, x_2' fixed which interchanges x_1 and x_1'. The two pairs are both concyclic and anticyclic.

In the case of the square this is intuitive, the two arcs being the two diagonals.

If we write

$$\pi = (x_1 - x_2)(x_1' - x_2')$$
$$\pi' = (x_1 - x_2')(x_1' - x_2)$$

then

$$\pi - \pi' = (x_1 - x_1')(x_2 - x_2')$$

and

$$\pi + \pi' = -p_{12}$$

The two ratios of π and π' are the cross-ratios of the two pairs. These are constants under all homographies.

We take as the fundamental constant of two pairs (each being ordered),

$$\kappa_{12} = \frac{(x_1 + x_1')(x_2 + x_2') - 2x_1x_1' - 2x_2x_2'}{(x_1 - x_1')(x_2 - x_2')}$$
$$= \frac{\pi + \pi'}{\pi' - \pi}$$

This is, in terms of the power p_{12},

$$p_{12}/\sqrt{p_{11}}\sqrt{p_{22}}$$

for when x_2, x_2' coincide with x_1, x_1', we have

$$p_{11} = (x_1 + x_1')^2 - 4x_1x_1' = (x_1 - x_1')^2$$

§ 28. **The Jacobian** – Two quadratics have one common normal pair. For if $\alpha x^2 + 2\beta x + \gamma$ is normal to $\alpha_i x^2 + 2\beta_i x + \gamma_i$, then

$$\alpha\gamma_i - 2\beta\beta_i + \gamma\alpha_i = 0,$$

and, eliminating α, β, γ, we have

$$\begin{vmatrix} x^2 & -x & 1 \\ \gamma_1 & \beta_1 & \alpha_1 \\ \gamma_2 & \beta_2 & \alpha_2 \end{vmatrix}$$

This is the Jacobian of the two forms.

For the two pairs x_i, x_i' the common normal is

$$\begin{vmatrix} x^2 & 2x & 1 \\ x_1x_1' & x_1 + x_1' & 1 \\ x_2x_2' & x_2 + x_2' & 1 \end{vmatrix}$$

For a canonical form of two pairs we take the Jacobian or common normal pair to be 0 and ∞. We have then

$$x_1 + x_1' = 0, \qquad x_2 + x_2' = 0$$

Thus the pairs are opposite vertices of a parallelogram.

To construct the Jacobian pair of x_1, x_1' and x_2, x_2' we take this case of the parallelogram. Here the arcs $x_1x_2x_1'$ and $x_1x_2'x_1'$ are equal, and the diagonal bisects the angle between these arcs. Hence in general we draw the circle which bisects the angle of the lune made by $x_1x_2x_1'$ and $x_1x_2'x_1'$, and the circle which bisects the angles of the lune made by $x_2x_1x_2'$ and $x_2x_1'x_2'$. These two circles meet in the Jacobian pair (fig. 22).

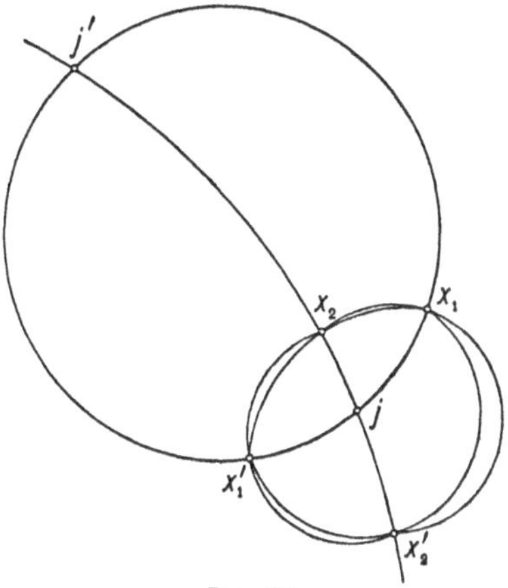

FIG. 22

Or since in the parallelogram case the circle on x_1 and x_1' as diameter is the exterior bisector of the angles made by $x_1x_2x_1'$ and $x_1x_2'x_1'$, in general the circles which bisect the exterior angles of the two lunes have the Jacobian pair as common inverse points.

The first construction fails when the given pair are on a circle and are not interlaced, for the two circles of construction coincide.

The second fails when the given pair are on a circle and are interlaced, for the same reason. One or other always holds.

Exercise 3 – The Jacobian pair for the pairs x_2, x_3 and x_1, ∞ are where the bisector of the angle $x_2x_1x_3$ (the internal bisector) meets the circle bisecting the angles which the line x_2, x_3, ∞ makes with the circle x_2, x_3, x_1.

§ 29. **The Vector** – The fundamental constant for the parallelogram ± 1, $\pm a$ is

$$\kappa_{12} = (2a^2 + 2)/4a$$

or

$$2\kappa_{12} = a + 1/a$$

or, if $a = \rho t = e^{\lambda + \iota\theta}$,

$$2\kappa_{12} = e^{\lambda}(\cos\theta + \iota\sin\theta) + e^{-\lambda}(\cos\theta - \iota\sin\theta)$$
$$= 2\cos\theta\cosh\lambda + 2\iota\sin\theta\sinh\lambda$$

or

$$\kappa_{12} = \cos(\theta - \iota\lambda)$$

We have thus attached to the two pairs x_1, x_1' and x_2, x_2' a vector $v=\theta-\iota\lambda$. If we change the order of one pair, v becomes $v+\pi$.

The involution set up by two points applies to any space on the points. When we select a plane Ω on the points, what is common to the plane and the involution is, in addition to the pairs in the plane, the pairs inverse as to the plane. These lie on the circle which is normal to the plane at the points. The proper way to realise a quadratic is then to think of a circle normal to the plane considered. It is enough to think of the space on one side of Ω, or if Ω is a sphere, of the inside of this sphere.

And this sphere may be regarded as a large sphere around one—bounding one's universe. This we develop in Chapter IX.

At the moment it is an apology for replacing a pair of points x_i, x_i' by a semicircle normal to the plane (or an arc normal to a sphere and lying inside it). This we call the arc x_i, x_i' when directed, or the arc q_i when not directed.

In the canonical case, when the arcs are from -1 to 1 and from $-a$ to a, we have

$$\cos(\theta-\iota\lambda)=\tfrac{1}{2}(a+1/a)$$

For the square, $a=\pm\iota$, and therefore

$$\theta-\iota\lambda=\pi/2$$

that is, $$\theta=\pi/2 \quad \text{and} \quad \lambda=0$$

The two arcs here cut at right angles. The complete circles to which the arcs belong cut twice at right angles. We call such circles or their arcs normal.

For a rectangle, a is a turn and $\cos(\theta-\iota\lambda)$ is the cosine of an angle. Therefore $\lambda=0$. Here the two arcs intersect.

Thus $\lambda=0$ characterises intersecting arcs or interlaced pairs of points.

For a rhombus, $a=\iota e^{\lambda}$ and

$$\cos(\theta-\iota\lambda)=\iota\sinh\lambda$$

that is, $\theta=\pi/2$. The two arcs are perpendicular, but do not intersect.

Thus $\theta=\pi/2$ characterises the rhombus.

For two pairs which, while concyclic, are not interlaced we have $a = e^{\lambda}$, and

$$\cos(\theta - \iota\lambda) = \cosh\lambda$$

Here then $\theta = 0$.

Since normal pairs of points give normal arcs, the common normal or Jacobian gives the arc normal to two arcs. In the canonical case this Jacobian arc is the line normal to Ω at 0. And we see from $a = e^{\lambda + \iota\theta}$ that θ is the angle made by the vertical plane on x, $-x$, 0, ∞ with that on 1, -1, 0, ∞.

Hence in general, when the Jacobian pair is j, j', the angle θ is the angle made by the sphere on the arcs x_2, x_2' ; j, j' with the sphere on the arcs x_1, x_1' ; j, j'.

Also in the parallelogram case e^{λ} is the ratio of the radii of the two arcs, in other words the hyperbolic distance of the arcs measured from Ω. And therefore in general λ is the hyperbolic distance of the arcs measured from Ω along the Jacobian. Thus, in passing from the arc x_1, x_1' to the arc x_2, x_2' along the common normal j, j', one turns through an angle θ, and goes a hyperbolic distance λ.

This vector $\theta - \iota\lambda$ is fundamental for any two circles in a Euclidean space *which are not interlaced.* For consider any two circles not on a sphere. For simplicity of expression take one of them as a line. There is an inversion which sends both the line and the circle into itself, the centre being where the line meets the plane of the circle. If the line does not pass through the circle, the inversion is hyperbolic and the sphere of fixed points is normal (twice orthogonal) to both the line and the circle.

If the line does pass through the circle the inversion is elliptic ; there is an extra pair. There is still a Jacobian circle twice orthogonal to both line and circle, but the whole treatment is different, and will not be given.

§ 30. **Theory of the Four-point** – The points ± 1, $\pm a$ can be paired also as

$$1, a \quad \text{and} \quad -1, -a$$

In this case the Jacobian is

$$\begin{vmatrix} x^2 & 2x & 1 \\ a & 1+a & 1 \\ a & -1-a & 1 \end{vmatrix}$$

or to a factor

$$x^2 - a$$

The Jacobian points are the square roots of a.

Or the points can be paired as

$$1, \ -a \quad \text{and} \quad -1, \ a$$

In this case the Jacobian is

$$x^2 + a$$

Thus the four-points as a whole have three Jacobian pairs,

$$2x, \qquad x^2 + a, \qquad x^2 - a$$

These are mutually normal.

The three Jacobian arcs are then mutually normal. They meet at an extra point on the normal to the plane at Ω, at a distance $|\sqrt{a}|$. If we invert with respect to this point, the plane becomes a sphere, and the three arcs become three diameters of it, mutually at right angles. The three Jacobian pairs have taken the canonical form of a regular octahedron.

Given the octahedron, the four-point is any point of the sphere, and its images in the three diagonals. It is thus four alternate corners of a rectangular box.

We notice here a distinction between the homographic theory of the four-point and the inversive theory. In the former there are three Jacobians. In the latter there are three Jacobians and an extra pair which we call the canonizant. In the latter there is also a counter four-point—the other four corners of the rectangular box.

So in the theory of the cubic, or three-point, it is primarily a question of points on a circle. On this circle is a counter-triad. But to complete the theory there are two extra points not on the circle.

We notice also that, just as the square with its centre and ∞ forms a configuration T_3 of three mutually normal pairs, all on the same footing, so the octahedron with its centre and ∞ forms a configuration T_4 of four mutually normal pairs. There is an elliptic inversion or extra pair which interchanges opposite vertices of the octahedron, and this, with the four pairs, gives a symmetrical configuration T_5 of five mutually normal pairs in four dimensions. And so on.

The configuration T_4, for example, is four normal pairs, 1, 2, 3, 4. These lie by twos on six circles such as 12. The circles 12 and 13 are normal—that is, cut twice at right angles. The six circles are by threes on four spheres such as 1 2 3 and these spheres are orthogonal or normal.

That is, we have 1 space,
on it 4 normal pairs,
6 circles,
4 spheres,
with 1 extra pair.

For the four-points x_i as a whole we have the six cross-ratios (§ 20):

$$-\pi_2/\pi_3, \quad -\pi_3/\pi_1, \quad -\pi_1/\pi_2$$
$$-\pi_3/\pi_2, \quad -\pi_1/\pi_3, \quad -\pi_2/\pi_1$$

and the three powers:

$$\pi_3 - \pi_2, \quad \pi_1 - \pi_3, \quad \pi_2 - \pi_1$$

The symmetric functions of these are the fundamental invariants of the four-point or of the quartic which gives them.

We have the three vectors v_i defined by

$$\cos v_1 = (\pi_2 - \pi_3)/\pi_1$$

or

$$\cos^2 v_1/2 = -\pi_3/\pi_1 \qquad \sin^2 v_1/2 = -\pi_2/\pi_1$$
$$\cos^2 v_2/2 = -\pi_1/\pi_2 \qquad \sin^2 v_2/2 = -\pi_3/\pi_2$$
$$\cos^2 v_3/2 = -\pi_2/\pi_3 \qquad \sin^2 v_3/2 = -\pi_1/\pi_3$$

The relations of the vectors are then

$$\sin^2 v_2/2 \cos^2 v_3/2 = 1$$
$$\sin^2 v_3/2 \cos^2 v_1/2 = 1$$
$$\sin^2 v_1/2 \cos^2 v_2/2 = 1$$

With four ordered points—say, $x_1, x_2, x_3, x_4, x_1 \ldots$—we associate a figure formed by four circles in a space, each touching the two next it. The points of contact are x_1, x_2, x_3, x_4. It is here convenient to take the arcs only.

The vector v_1 for the arcs x_4, x_1 and x_2, x_3 is given by

$$\cos v_1 = (\pi_2 - \pi_3)/\pi_1$$

whence

$$\cos^2 v_1/2 = -\pi_3/\pi_1, \quad \sin^2 v_1/2 = -\pi_2/\pi_1$$

The vector for the arcs x_1, x_2 and x_3, x_4 is not v_3, but $v_3 + \pi$. Calling this $v_3{}'$,

$$\cos v_3{}' = (\pi_2 - \pi_1)/\pi_3$$

whence

$$\cos^2 v_3{}'/2 = -\pi_1/\pi_3, \quad \sin^2 v_3{}'/2 = -\pi_2/\pi_3$$

Hence

$$\cos^2 v_1/2 \, \cos^2 v_3{}'/2 = 1$$

Exercise 4 – Show by a special case that for four arcs with continuity of direction

$$\cos v_1/2 \, \cos v'_3/2 = 1$$

Exercise 5 – If

$$\cos \alpha \sin \beta = 1$$

and

$$\cos \beta \sin \gamma = 1$$

then

$$\cos \gamma \sin \alpha = \pm 1$$

§ 31. The Ordered Six-point

– With four ordered points $a\,b\,c\,d$ are associated the pairs a, c and b, d.

The Jacobian of these pairs is given by

$$[abcd] \equiv \frac{(x-a)(x-b)}{a-b} + \frac{(x-b)(x-c)}{b-c}$$
$$+ \frac{(x-c)(x-d)}{c-d} + \frac{(x-d)(x-a)}{d-a}$$

For this cyclical expression is

$$\frac{(x-b)^2(a-c)}{(a-b)(b-c)} + \frac{(x-d)^2(c-a)}{(c-d)(d-a)}$$

and is therefore normal to

$$(x-b)(x-d)$$

And similarly it is normal to

$$(x-a)(x-c)$$

Thus $[abcd]$ is a covariant of the ordered four-point.

For any number of ordered points we have a similar quadratic covariant. Let us take the case of six points, $a\,b\,c\,d\,e\,f$, which we may call a hexagon. The quadratic is the Kirkman quadratic of the hexagon.*

---

* If we restrict numbers to be real, the theory of this section becomes the theory of six points on a conic. The quadratics here are named in conformity with that theory (Salmon, *Conic Sections*, Appendix D; Baker, *Principles of Geometry*, vol. 2). For extensions of this section see a note on the Celestial Sphere, *Amer. Journal*, vol. 54, 1932.

Let us write the six points as

$$\begin{matrix} a & b & c \\ f & e & d \end{matrix}$$

We have then set up a homography, in which c, d; b, e; a, f shall correspond. This is

$$\begin{vmatrix} xy & x & y & 1 \\ af & a & f & 1 \\ be & b & e & 1 \\ cd & c & d & 1 \end{vmatrix}$$

It has a fixed pair, or Pascal quadratic, which we denote by $\begin{matrix} a\,b\,c \\ f\,e\,d \end{matrix}$. With these as 0 and ∞ it has the canonical form $y=\kappa x$. We may then take the six points as $a, b, c, \kappa c, \kappa b, \kappa a$.

The Kirkman quadratic is

$$\frac{(x-a)(x-b)}{a-b}+\frac{(x-b)(x-c)}{b-c}+\frac{(x-c)(x-\kappa c)}{c(1-\kappa)}+$$
$$+\frac{(x-\kappa c)(x-\kappa b)}{\kappa(c-b)}+\frac{(x-\kappa b)(x-\kappa a)}{\kappa(b-a)}+\frac{(x-\kappa a)(x-a)}{a(\kappa-1)}$$

The coefficient of x is 0.

Therefore the Kirkman quadratic $[abcdef]$ is normal to the pair $\begin{matrix} a\,b\,c \\ f\,e\,d \end{matrix}$. This latter is a Pascal quadratic. It is associated with the ordering $a\,e\,c\,f\,b\,d$.

Thus the Kirkman $[abcdef]$ is normal to the Pascals

$$\begin{matrix} a\,b\,c \\ f\,e\,d \end{matrix} \qquad \begin{matrix} b\,c\,d \\ a\,f\,e \end{matrix} \qquad \begin{matrix} c\,d\,e \\ b\,a\,f \end{matrix}$$

and the Pascal

$$\begin{matrix} a\,b\,c \\ f\,e\,d \end{matrix} \equiv \begin{matrix} b\,c\,a \\ e\,d\,f \end{matrix} \equiv \begin{matrix} c\,a\,b \\ d\,f\,e \end{matrix}$$

is normal to the Kirkmans $[abcdef]$, $[bcafde]$, $[cabefd]$.

The above quadratic for a polygon, $[abc\ldots]$, is obtained by taking the Jacobian of $(x-a)^2$ and $(x-b)^2$ and dividing by the bilinear invariant $(a-b)^2$.

It equally applies to a cycle of quadratics $q_1q_2\ldots q_n$ in the form

$$j_{12}/q_{12}+j_{23}/q_{23}+\ldots+j_{n1}/q_{n1}$$

where j_{12} is the Jacobian and q_{12} the bilinear invariant.

§ 32. **The Complete System** – For n quadratics $q_1 \ldots q_n$, where $q_i = a_i\bar{x} + 2\beta_i x + \gamma_i$ the complete system of invariants (Gordan, *Invarianten-theorie*; Grace and Young, *Algebra of Invariants*) is the n discriminants

$$q_{11}, \quad q_{22} \ldots$$

the $\binom{n}{2}$ forms

$$q_{12}, \quad q_{13}, \quad q_{23} \ldots$$

and the $\binom{n}{3}$ invariants which express that any three are in involution, such as

$$\begin{vmatrix} a_1 & \beta_1 & \gamma_1 \\ a_2 & \beta_2 & \gamma_2 \\ a_3 & \beta_3 & \gamma_3 \end{vmatrix}$$

and the complete system of covariants is the n quadratics and the $\binom{n}{2}$ Jacobians.

For n homographic forms $H_1 \ldots H_n$ we may replace H_i by the quadratic q_i of fixed points and the invariant $\beta_i - \gamma_i$.

The complete system (remarks Professor Coble) will be the complete system for the q_i and the invariants $\beta_i - \gamma_i$.

CHAPTER VI

THE INVERSIVE GROUP OF THE PLANE

§ 33. **Fixed Points** – A homography has two fixed points given by $\alpha f^2 + (\beta+\gamma)f + \delta = 0$. In terms of these, when distinct, it is

$$\frac{x-f_1}{x-f_2} = \kappa \frac{y-f_1}{y-f_2}$$

If the fixed points be 0 and ∞, so that $\alpha=\delta=0$, the homography takes the canonical form

$$x = \kappa y$$

When the fixed points coincide, the canonical form is found by making $\alpha=0$, $\beta+\gamma=0$, so that both values of f are ∞. It is then $x=y+b$.

An antigraphy may or may not have fixed points. We ask here for the points such that

$$\alpha x\bar{y} + \beta x + \gamma \bar{y} + \delta = 0$$

and

$$\alpha y\bar{x} + \beta y + \gamma \bar{x} + \delta = 0$$

Writing the conjugate of the second,

$$\bar{\alpha} x\bar{y} + \bar{\beta}\bar{y} + \bar{\gamma} x + \bar{\delta} = 0$$

we have two equations in x and $\bar{y}$, and a quadratic for x. Let the roots be f_1 and f_2. Then the antigraphy may be either

$$\frac{x-f_1}{x-f_2} = \kappa \frac{\bar{y}-\bar{f}_1}{\bar{y}-\bar{f}_2}$$

or

$$\frac{x-f_1}{x-f_2} = \kappa \frac{\bar{y}-\bar{f}_2}{\bar{y}-\bar{f}_1}$$

There will be either two fixed points or an interchanging pair.

With these two points as 0 and ∞, we have either

$$x = \kappa\bar{y}$$

or

$$x\bar{y} = \kappa$$

In the former case, by writing $x = \tau x'$, we can make κ a positive, p.

In the latter, by writing $x = \rho x'$, we can make κ a turn, t. Thus we take for the canonical forms of an antigraphy

$$x = p\bar{y}$$
$$x\bar{y} = t$$

When the two points coincide, the canonical form is (§ 15)

$$x = \bar{y} + p$$

From these canonical forms we see that any antigraphy can be the product of three inversions. Thus $x = p\bar{y}$ is the product of $x = \bar{y}$ and $x = py$, and $x = py$ is the product of concentric inversions. And $x\bar{y} = t$ is the product of $x\bar{y} = 1$ and $x = ty$, which again is three inversions.

The group formed by all antigraphies is then the inversive group, formed by all inversions of a sphere into itself. An odd number of inversions gives an antigraphy, an even number a homography. For the general homography we see from $x = pty$ that four inversions are necessary, two which send x into py, and two which send x into ty.

§ 34. **Invariants of a Homography** – As generators of the inversive group we may take

$$\begin{aligned}
&(1) \quad x = x' + b \\
&(2) \quad x = ax' \\
&(3) \quad x = 1/x' \\
&(4) \quad x = \bar{x}'
\end{aligned}$$

If we omit (4) we have the generators of the homographic group, the subgroup formed by all homographies.

A function of points x_i which becomes to a factor the same function of x_i' under all these generators is an invariant. And a function of points x_i and a variable x which becomes to a factor the same function of x_i' and x' is a covariant. For the discovery or the formal proof of invariance or covariance one may apply the generators. Thus one may say:

under (1) $x_1 - x_2$ is constant;
under (2) $x_1 - x_2$ is an invariant, and $(x_1 - x_2)/(x_1 - x_3)$ is a constant;
under (3) this is an invariant, but

$$\frac{(x_1 - x_2)(x_3 - x_4)}{(x_1 - x_3)(x_2 - x_4)}$$

is a constant.

This cross-ratio is then constant under (1), (2), (3), that is under H.

But in the simple cases considered in this book we shall not as a rule supply the formal proofs of invariance.

The points of a sphere S are named by assigning to three points the numbers 1, ∞, 0. The points of a second overlying sphere S′ are similarly named. Let x and x' be overlying points. Then there is a definite homography, say

$$x' = (ax + b)/(cx + d)$$

the points 0 and ∞ of S coinciding with the points b/d and a/c of S′; and the points 0 and ∞ of S′ coinciding with the points $-b/a$ and $-d/c$ of S.

Let us take on S a set of points given by $f(x, \alpha) = 0$, or a curve given by a self-conjugate equation $f(x, \bar{x}, \alpha, \bar{\alpha}) = 0$, or in fact any object algebraically given. We have overlying it on S′ the same object differently expressed, the new expression being obtained by writing, for x,

$$\frac{b - dx'}{cx' - a}$$

A property of the object itself will of course be invariant however expressed. It will be independent of the reference scheme adopted.

Consider the invariants of a homography H. The form here is

$$\alpha xy + \beta x + \gamma y + \delta$$

It is of period 2 (a polarity) if $\beta = \gamma$.

Hence $g_1 \equiv \beta - \gamma$ is an invariant.
If we write

$$x = x' + b, \qquad y = y' + b$$

we get

$$\alpha' x' y' + \beta' x' + \gamma' y' + \delta'$$

where

$$\alpha' = \alpha, \qquad \beta' = \alpha b + \beta, \qquad \gamma' = \alpha b + \gamma$$

Thus, under (1), $\beta - \gamma$ is a constant.

Exercise 1 – Verify that under (2) and (3) it is an invariant.

Again the form has factors when $\alpha\delta = \beta\gamma$.
Hence $g_2 \equiv \alpha\delta - \beta\gamma$ is an invariant.

For the canonical form $x = \kappa y$,

$$g_1 = 1 + \kappa, \qquad g_2 = \kappa$$

Hence

$$g_1^2/g_2 = \kappa + 1/\kappa + 2$$

Of the two reciprocal values of κ, one belongs to H, the other to H^{-1}.

Exercise 2 – The fixed points coincide when $g_1^2/g_2 = 4$.

Exercise 3 – A homography is of period 3 when $g_1^2/g_2 = 3$; of period 4 when $g_1^2/g_2 = 2$; of period n when

$$g_1^2/g_2 = 4 \cos^2 (m\pi/n)$$

where m is prime to n.

Homographies are classified by the value of κ or of g_1^2/g_2. When κ is a real, the homography is hyperbolic; when κ is a turn, the homography is elliptic; when κ is a general number the homography is loxodromic.

Exercise 4 – The homography given by inversions in two circles is elliptic when the two circles are intersecting, hyperbolic when the two circles are non-intersecting, and parabolic when the circles are touching.

§ 35. **Composition of Homographies** – Two polarities Q_1, Q_2 have in general a common pair of points, namely, the Jacobian of their fixed points. Taking the common pair as 0 and ∞, Q_1 and Q_2 become

$$xy = \kappa_1$$
$$xy = \kappa_2$$

The product $Q_1 Q_2$ is then

$$y = \kappa_2 x / \kappa_1$$

the homography H with fixed points 0, ∞.

Conversely,

$$y = \kappa x$$

can be broken up into

$$xy = \kappa_1$$
$$xy = \kappa\kappa_1$$

Thus a homography H is resolved into two polarities, whose fixed points are harmonic to those of H, and one pair of fixed points can be taken as any pair harmonic to the fixed points of H.

Given then three homographies H_1, H_2, H_3, whose product is I (identity), their fixed points have three Jacobians J_1, J_2, J_3. These are the fixed points of three polarities Q_i such that

$$H_1 = Q_2Q_3, \qquad H_2 = Q_3Q_1, \qquad H_3 = Q_1Q_2$$

§ **36. Invariants of an Antigraphy** – The product of two antigraphies,

$$A_1 : \quad \alpha_1 x\bar{y} + \beta_1 x + \gamma_1\bar{y} + \delta_1$$
$$A_2 : \quad \alpha_2 y\bar{z} + \beta_2 y + \gamma_2\bar{z} + \delta_2$$

is, writing for the second its conjugate and eliminating $\bar{y}$, the homography

$$A_1A_2 : \quad \begin{vmatrix} \alpha_1 x + \gamma_1 & \beta_1 x + \delta_1 \\ \bar{\alpha}_2 z + \bar{\beta}_2 & \bar{\gamma}_2 z + \bar{\delta}_2 \end{vmatrix}$$

In particular the square of an antigraphy A is

$$A^2 : \quad (\alpha\bar{\gamma} - \beta\bar{\alpha})xz + (\alpha\bar{\delta} - \beta\bar{\beta})x + (\gamma\bar{\gamma} - \delta\bar{\alpha})z + \gamma\bar{\delta} - \delta\bar{\beta}$$

The simplest invariant of A is the g_1 of A^2—that is,

$$g_1 \equiv \alpha\bar{\delta} + \delta\bar{\alpha} - \beta\bar{\beta} - \gamma\bar{\gamma}$$

Its vanishing implies that $A^4 \equiv I$. The points of the plane are arranged cyclically in sets of four.

If, for instance,

$$A \text{ is } x\bar{y} - \iota$$

then

$$A^2 \text{ is } x + y$$

For any point x the cycle x, xA, xA^2, xA^3 is

$$x, \qquad -\iota/\bar{x}, \qquad -x, \qquad \iota/\bar{x}$$

the successive vertices of a rhombus.

A second invariant of A is found from $\alpha\delta - \beta\gamma$. For this is the condition that A has factors. Under (1), (2), and (3) of § 34 it is an invariant—that is, it reappears to a factor. But under (4) it becomes its conjugate. The invariant is

$$g_2 \equiv (\alpha\delta - \beta\gamma)(\bar{\alpha}\bar{\delta} - \bar{\beta}\bar{\gamma})$$

The fixed points of the homography A^2 are given by

$$(\alpha\bar{\gamma} - \beta\bar{\alpha})f^2 + (\alpha\bar{\delta} - \beta\bar{\beta} + \gamma\bar{\gamma} - \delta\bar{\alpha})f + \gamma\bar{\delta} - \delta\bar{\beta}$$

These may be fixed individually under A, or interchanged, or coincident. The discriminant of the quadratic is

$$4(\alpha\bar{\gamma} - \beta\bar{\alpha})(\gamma\bar{\delta} - \delta\bar{\beta}) - (\alpha\bar{\delta} - \beta\bar{\beta} + \gamma\bar{\gamma} - \delta\bar{\alpha})^2$$

This is

$$4(\alpha\delta - \beta\gamma)(\bar{\alpha}\bar{\delta} - \bar{\beta}\bar{\gamma}) - (\alpha\bar{\delta} + \delta\bar{\alpha} - \beta\bar{\beta} - \gamma\bar{\gamma})^2$$

or

$$4g_2 - g_1^2$$

The quadratic, being inversively attached to the antigraphy, is called a covariant.

Exercise 5 – Its identical vanishing is the condition that A be an inversion.

For the canonical form $x = p\bar{y}$, we have

$$\alpha = \delta = 0, \qquad \beta = 1, \qquad \gamma = -p, \qquad g_1 = -1 - p^2, \qquad g_2 = p^2$$

Whereas for the form $x\bar{y} = t$ we have

$$g_1 = -(t + \iota/t), \qquad g_2 = 1$$

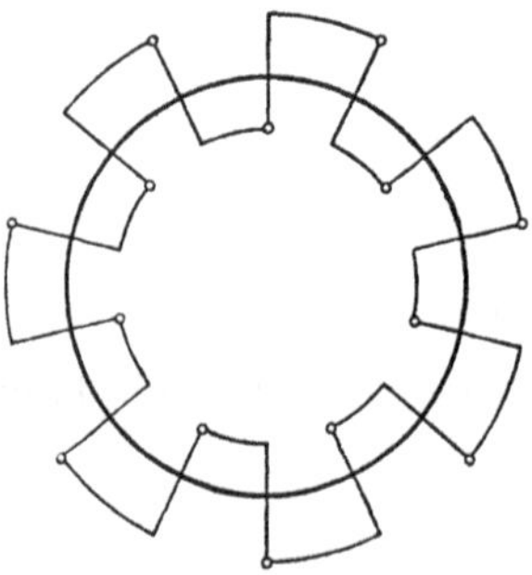

FIG. 23

Thus in the former $g_1^2/g_2 > 4$, and in the latter $g_1^2/g_2 < 4$.

It will be noticed that, whereas in a homography the constant g_1^2/g_2 is any number, in an antigraphy it is real.

An antigraphy of the form $x\bar{y} = t$ will be periodic when $x = t^2 z$ is periodic—that is, when t is a root of unity. Fig. 23 illustrates this, the period being here 14.

§ 37. **The Canonical Form** – All antigraphies are included in

(1) $$(\lambda C) \equiv \lambda_0 C_0 + \lambda_1 C_1 + \lambda_2 C_2 + \lambda_3 C_3$$

(§ 26) when the λ_i are no longer real. For this is

$$\alpha x\bar{y} + \beta x + \gamma\bar{y} + \delta$$

where

$$\iota\lambda_0 + \lambda_2 = \alpha, \qquad \lambda_3 + \iota\lambda_1 = \beta$$
$$\iota\lambda_0 - \lambda_2 = \delta, \qquad \lambda_3 - \iota\lambda_1 = \gamma$$

Expressed in terms of the λ_i the invariants take the symmetrical forms

$$g_1 = 2(\lambda_0\bar{\lambda}_0 + \lambda_1\bar{\lambda}_1 + \lambda_2\bar{\lambda}_2 + \lambda_3\bar{\lambda}_3)$$
$$g_2 = (\lambda_0^2 + \lambda_1^2 + \lambda_2^2 + \lambda_3^2)(\bar{\lambda}_0^2 + \bar{\lambda}_1^2 + \bar{\lambda}_2^2 + \bar{\lambda}_3^2)$$

Similarly, if we write

$$h_0 = y - x$$
$$h_1 = \iota(x + y)$$
$$h_2 = \iota(xy - 1)$$
$$h_3 = xy + 1$$

where

$$h_0^2 + h_1^2 + h_2^2 + h_3^2 = 0$$

any homography is

(2) $$(ah) \equiv a_0h_0 + a_1h_1 + a_2h_2 + a_3h_3$$

where

$$\alpha = a_3 + \iota a_2$$
$$\delta = a_3 - \iota a_2$$
$$\beta = \iota a_1 - a_0$$
$$\gamma = \iota a_1 + a_0$$

so that

$$\alpha\delta - \beta\gamma = a_0^2 + a_1^2 + a_2^2 + a_3^2$$

For two homographies in the form (ah), (βh) we have

$$\alpha\delta' + \alpha'\delta = (a_3 + \iota a_2)(\beta_3 - \iota\beta_2) + (a_3 - \iota a_2)(\beta_3 + \iota\beta_2)$$
$$= 2(a_3\beta_3 + a_2\beta_2)$$

and

$$\beta\gamma' + \beta'\gamma = -2(a_0\beta_0 + a_1\beta_1)$$

so that the orthogonal invariant is

(3) $$(a\beta) \equiv a_0\beta_0 + a_1\beta_1 + a_2\beta_2 + a_3\beta_3$$

For the fundamental four h_i the six such invariants h_{ij} are all 0.

We define as the product of the homographic forms

$$\alpha xy + \beta x + \gamma y + \delta$$

and

$$\alpha' yz + \beta' y + \gamma' z + \delta'$$

in this order, the form

$$\begin{vmatrix} \alpha x+\gamma & \beta x+\delta \\ \alpha' z+\beta' & \gamma' z+\delta' \end{vmatrix}$$

or

$$(\alpha\gamma'-\beta\alpha')xz+(\alpha\delta'-\beta\beta')x+(\gamma\gamma'-\delta\alpha')z+\gamma\delta'-\delta\beta'$$

This product is so chosen that its discriminant is the product of the discriminants.

We have then *

$$h_0{}^2=\begin{pmatrix} 0 & -1 & 1 & 0 \\ 0 & -1 & 1 & 0 \end{pmatrix}=h_0$$

$$h_1{}^2=\begin{pmatrix} 0 & \iota & \iota & 0 \\ 0 & \iota & \iota & 0 \end{pmatrix}=-h_0$$

$$h_2{}^2=\begin{pmatrix} \iota & 0 & 0 & -\iota \\ \iota & 0 & 0 & -\iota \end{pmatrix}=-h_0$$

$$h_3{}^2=\begin{pmatrix} 1 & 0 & 0 & 1 \\ 1 & 0 & 0 & 1 \end{pmatrix}=-h_0$$

$$h_0h_1=h_1h_0=h_0$$
$$h_0h_2=h_2h_0=h_0$$
$$h_0h_3=h_3h_0=h_0$$

$$h_1h_2=\begin{pmatrix} 0 & \iota & \iota & 0 \\ \iota & 0 & 0 & -\iota \end{pmatrix}=h_3=-h_2h_1$$

$$h_2h_3=h_1=-h_3h_2$$
$$h_3h_1=h_2=-h_1h_3$$

whence

$$h_1h_2h_3=h_2h_3h_1=h_3h_1h_2=-h_0$$
$$h_3h_2h_1=h_2h_1h_3=h_1h_3h_2=h_0$$

Hence the product

$$\begin{aligned}(\alpha_0h_0+\alpha_1h_1+\alpha_2h_2+\alpha_3h_3)&(\beta_0h_0+\beta_1h_1+\beta_2h_2+\beta_3h_3)\\ &=\alpha_0(\beta_0h_0+\beta_1h_1+\beta_2h_2+\beta_3h_3)\\ &+\alpha_1(\beta_0h_1-\beta_1h_0+\beta_2h_3-\beta_3h_2)\\ &+\alpha_2(\beta_0h_2-\beta_1h_3-\beta_2h_0+\beta_3h_1)\\ &+\alpha_3(\beta_0h_3+\beta_1h_2-\beta_2h_1-\beta_3h_0)\\ &=(\alpha_0\beta_0-\alpha_1\beta_1-\alpha_2\beta_2-\alpha_3\beta_3)h_0\\ &+(\alpha_0\beta_1+\alpha_1\beta_0+\alpha_2\beta_3-\alpha_3\beta_2)h_1\\ &+(\alpha_0\beta_2+\alpha_2\beta_0-\alpha_1\beta_3+\alpha_3\beta_1)h_2\\ &+(\alpha_0\beta_3+\alpha_3\beta_0+\alpha_1\beta_2-\alpha_2\beta_1)h_3\end{aligned}$$

* It must be observed that notations such as $h_0{}^2$ are used here in two senses. The square of the expression $y-x$ is of course $(y-x)^2$, but the square of the operation given by $y-x$ is given by $y-x$.

We have here an introduction to quaternions. We are dealing with four unitary operations h_0, h_1, h_2, h_3, and we have a rule for the product of two quaternions (αh) and (βh) as a third quaternion.

§ 38. **The Determinant of Powers** – Between five linear forms, either

$$f_i \equiv \alpha_i x\bar{y} + \beta_i x + \gamma_i \bar{y} + \delta_i$$

or

$$f_i \equiv \alpha_i xy + \beta_i x + \gamma_i y + \delta_i$$

there is the linear relation

$$| \, f_i, \quad \alpha_i, \quad \beta_i, \quad \gamma_i, \quad \delta_i \, | = 0$$

say

$$\mu_1 f_1 + \mu_2 f_2 + \mu_3 f_3 + \mu_4 f_4 + \mu_5 f_5 = 0$$

Hence, if we denote the power of f_i and f_j by f_{ij},

$$\mu_1 f_{i1} + \mu_2 f_{i2} + \ . \ . \ . \ = 0$$

Thus between the powers and discriminants there is the relation

$$(1) \qquad | \, f_{i1} \quad f_{i2} \quad f_{i3} \quad f_{i4} \quad f_{i5} \, | = 0$$

When f_5 is degenerate, that is when $f_{55} = 0$, f_{i5} is merely f_i itself with the singular points of f_5 as new variables. Thus there is the quadratic relation on four forms,

$$(2) \qquad \begin{vmatrix} f_{11} & f_{12} & f_{13} & f_{14} & f_1 \\ f_{21} & f_{22} & f_{23} & f_{24} & f_2 \\ f_{31} & f_{32} & f_{33} & f_{34} & f_3 \\ f_{41} & f_{42} & f_{43} & f_{44} & f_4 \\ f_1 & f_2 & f_3 & f_4 & 0 \end{vmatrix} = 0$$

Suppose, as a case of (1), that we have five circles in a plane. Then between their powers there is the relation (1), always.

If we have the relation for four circles,

$$| \, f_{i1} \quad f_{i2} \quad f_{i3} \quad f_{i4} \, | = 0$$

then there is a circle f_5 (or a bilinear curve) normal to the four. For, if we take $f_{51} = 0$, $f_{52} = 0$, $f_{53} = 0$, then $f_{54} = 0$.

If we have for three circles the relation

$$| \, f_{i1} \quad f_{i2} \quad f_{i3} \, | = 0$$

then there are more than one circle normal to the three. The three then meet at a point.

If we have for two circles the relation

$$| \; f_{i1} \quad f_{i2} \; | = 0$$

the circles touch.

There is similarly for two sets of five forms, f_i and f_i' the general relation of powers

$$| \; f_{i1'} \quad f_{i2'} \quad f_{i3'} \quad f_{i4'} \quad f_{i5'} \; | = 0$$

This kind of fact is fundamental. It applies to any things two of which have a bilinear invariant, for example to curves of a plane of the same degree (in x or in $\bar{x}$ severally), to spheres, or to surfaces of the same inversive degree.

Let us indicate a way of writing the bilinear invariant of two curves or two surfaces.

We take in a Euclidean plane or space fixed points x_i and a variable point x. We denote the power or squared distance between x and x_i by p_i, and that between x_i and x_j by p_{ij}.

A spread is given by a form

$$\sum^{m} \lambda_i p_i^{\,n} \tag{3}$$

where the reals λ_i are the homogeneous co-ordinates of the spread with reference to the fixed points x_i. The number m is the number of co-ordinates; for enough points x_i it becomes the number of co-ordinates for the spread considered, independent of the x_i. This number was assigned (§ 26). Lastly, we call n the inversive degree of the spread. Thus, when the space is one-way (a line or circle), the spread is an even number of points; when the space is two-way (a plane or sphere) the spread is an algebraic curve; when the space is three-way the spread is an algebraic surface. Taking the planar case,

$$\sum^{4} \lambda_i p_i \text{ is any bilinear curve}$$

$$\sum^{9} \lambda_i p_i^2 \text{ is any biquadratic curve}$$

and so on.

The essential fact is that, under an inversion (x_0, κ), a power p_{12} becomes p_{12}' where (§ 23)

$$p_{12}' = \kappa^2 p_{12}/p_{01}p_{02}$$

Applied to (3) this shows that the spread is inversively related to the points x_i.

For

$$\Sigma\lambda_i p_i^n \quad \text{and} \quad \Sigma\mu_i p_i^n$$

the expression

$$\tfrac{1}{2}\Sigma(\lambda_i\mu_j + \lambda_j\mu_i)p_{ij}^n \tag{4}$$

is an invariant. For under an inversion (x_0, κ) the term $\lambda_1 p_1^n$ becomes $\lambda_1' p_1'^n$ where

$$p_1' = \kappa^2 p_1/p_{01}p_0$$

whence

$$\lambda_1' = \lambda_1 p_{01}^n p_0^n/\kappa^{2n}$$

Similarly for $\mu_2 p_2^n$,

$$\mu_2' = \mu_2 p_{02}^n p_0^n/\kappa^{2n}$$

while

$$p_{12}' = \kappa^2 p_{12}/p_{01}p_{02}$$

Hence

$$\lambda_1'\mu_2' p_{12}'^n = \lambda_1\mu_2 p_{12}^n \times p_0^{2n}/\kappa^{4n-2}$$

Thus two spreads of the same degree and of like dimension have a bilinear invariant.

To apply this to obtain covariants we polarise the spreads; that is we write, for p_i^n, $p_i^{n-r}q_i^r$, where q_i is the power from a second point y to x_i. We then form the invariant for, say, the forms in p_i,

$$\Sigma\lambda_i p_i^{n-r}q_i^r \quad \text{and} \quad \Sigma\mu_i p_i^{n-r}q_i^r$$

giving

$$\tfrac{1}{2}\Sigma(\lambda_i\mu_j + \lambda_j\mu_i)p_{ij}^{n-r}q_i^r q_j^r \tag{5}$$

a series of quadratic covariants of the two forms, called *transvectants*. With one of these and the given forms we form cubic transvectants; and so on. In particular, when the two forms coincide, we have from (4) the quadratic invariant

$$\Sigma\lambda_i\lambda_j p_{ij}^n \tag{6}$$

and from (5) a series of transvectants, quadratic in the coefficients.

Consider the case of six spheres. Since there is an identity $\Sigma\lambda_i S_i = 0$, the determinant of powers $|p_{ij}|$ is 0. If five spheres be orthogonal to the sixth, then $p_{i6} = 0$ ($i = 1 \ldots 5$). Hence the five-rowed determinant $|p_{ij}| = 0$. If in particular the five spheres are double points, then we have Cayley's condition that five points x_i be on a sphere. Let x_5 move to ∞. We have Cayley's condition that four points be on a plane,

$$\begin{vmatrix} 0 & p_{12} & p_{13} & p_{14} & 1 \\ p_{21} & 0 & p_{23} & p_{24} & 1 \\ p_{31} & p_{32} & 0 & p_{34} & 1 \\ p_{41} & p_{42} & p_{43} & 0 & 1 \\ 1 & 1 & 1 & 1 & 0 \end{vmatrix} = 0$$

If we write $p_{i4} = (\xi_i + \rho)^2$, where $i = 1$, 2, or 3, we have an equation for the radius ρ of the circle which touches externally the three circles (x_i, ξ_i). This equation is

$$\begin{vmatrix} 0 & p_{12} & p_{13} & \xi_1^2 + 2\rho\xi_1 & 1 \\ p_{21} & 0 & p_{23} & \xi_2^2 + 2\rho\xi_2 & 1 \\ p_{31} & p_{32} & 0 & \xi_3^2 + 2\rho\xi_3 & 1 \\ \xi_1^2 + 2\rho\xi_1 & \xi_2^2 + 2\rho\xi_2 & \xi_3^2 + 2\rho\xi_3 & -2\rho^2 & 1 \\ 1 & 1 & 1 & 1 & 0 \end{vmatrix} = 0$$

The three circles will touch a line when the coefficient of ρ^2 is 0, that is when

$$2\begin{vmatrix} 0 & p_{12} & p_{13} & \xi_1 & 1 \\ p_{21} & 0 & p_{23} & \xi_2 & 1 \\ p_{31} & p_{32} & 0 & \xi_3 & 1 \\ \xi_1 & \xi_2 & \xi_3 & 0 & 1 \\ 1 & 1 & 1 & 1 & 0 \end{vmatrix} - \begin{vmatrix} 0 & p_{12} & p_{13} & 1 \\ p_{21} & 0 & p_{23} & 1 \\ p_{31} & p_{32} & 0 & 1 \\ 1 & 1 & 1 & 0 \end{vmatrix} = 0$$

The ξ_i are the distances from x_i to any line. Thus we have the fundamental quadratic relation between the mutual distances of three points, and their distances from a line. An elementary proof is easy; but the argument as stated is general.

The argument of this section is readily extended to elliptic or hyperbolic spaces, replacing each power p by $(2 \sin \theta/2)^2$ or $(2 \sinh \delta/2)^2$.

CHAPTER VII

FINITE INVERSIVE GROUPS

§ 39. The Inversive Group of the Three-point – We consider the inversions of three points into themselves. There are four, namely, the one which leaves each point fixed, and the three which leave one point fixed and interchange the other two. The first is inversion in the circumcircle; the fixed circles of the other three inversions are called the *Apollonian circles* of the three-point. These four operations, with their products, are the inversive group of the three-point. It is a finite group.

As a canonical case we take the points to be 1, ω, ω^2. The fixed circles are then

$$x\bar{x}=1$$
$$\bar{x}=x$$
$$\bar{x}=\omega x$$
$$\bar{x}=\omega^2 x$$

that is, the circumcircle and three lines which meet at angles of $2\pi/3$. Hence in general the Apollonian circles always meet at two points h and h', at angles $2\pi/3$.

The inversions are

$$x\bar{y}=1$$
$$\bar{y}=x$$
$$\bar{y}=\omega x$$
$$\bar{y}=\omega^2 x$$

The twelve products of pairs of these give three polarities of period two,

$$xy=1$$
$$xy=\omega$$
$$xy=\omega^2$$

and two homographies of period three,

$$y=\omega x$$
$$y=\omega^2 x$$

From these and the inversions we get two reciprocal antigraphies of period six,

$$x\bar{y}=\omega$$
$$x\bar{y}=\omega^2$$

These, with identity, are the complete group of twelve operations, G_{12}.

This is clear without forming the products, for the plane is divided by the four circles into twelve *fundamental* regions. A selected region I passes by the four inversions into four other regions 0, 1, 2, 3. By repeated inversions the region I can pass only into one of the twelve regions (fig. 24). Thus there are twelve operations.

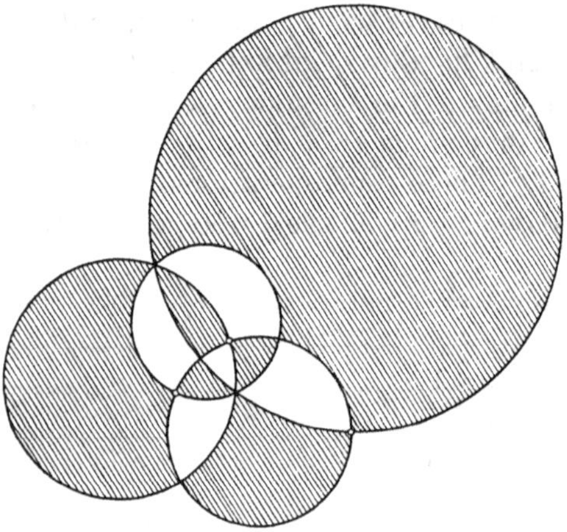

FIG. 24

The positive operations, those with an even number of inversions, are the homographic group of the three-point. This is a G_6. This group is the group of all permutations of three letters; for we are sending 1, 2, 3 into a new order. We can send 1, 2, 3 into 1, 3, 2. This is the polarity with fixed point x_1, interchanging x_2 and x_3. We can send 1, 2, 3 into 2, 3, 1. This is an operation H, whose square sends 1, 2, 3 into 3, 1, 2, and whose cube sends 1, 2, 3 into 1, 2, 3. It is a homography of period three.

The permutation group is denoted by

I, (23), (31), (12), (123), (321),

that is, by the cyclic interchanges or cycles, here of period two and three respectively. Identity is of period one. So it is convenient to denote the antigraphies in G_{12} by their cycles, $\bar{1}\,\bar{2}\,\bar{3}$, $\overline{23}$, $\overline{31}$, $\overline{12}$, $\overline{123}$, $\overline{321}$.

In fig. 24, if we distinguish the positive and negative

regions, by the signs + and −, or by shading one of the sets, then a positive and a negative region with a common edge will form a fundamental region for the group G_6. Any one region is a fundamental region for G_{12}.

Either the figure or the equations show that there are inversively attached to the three-point a two-point (in the canonical form the points 0, ∞) and another three-point (the points -1, $-\omega$, $-\omega^2$). These are respectively the *Hessian* and the *counter-triad.*

Exercise 1 – The three-point and its counter-triad form an inverse of a regular hexagon. The group of transformations of the hexagon into itself is found by adjoining to G_{12} the transformation

$$y+x=0$$

or the transformation

$$x\bar{y}+1=0$$

It is then a G_{24}.

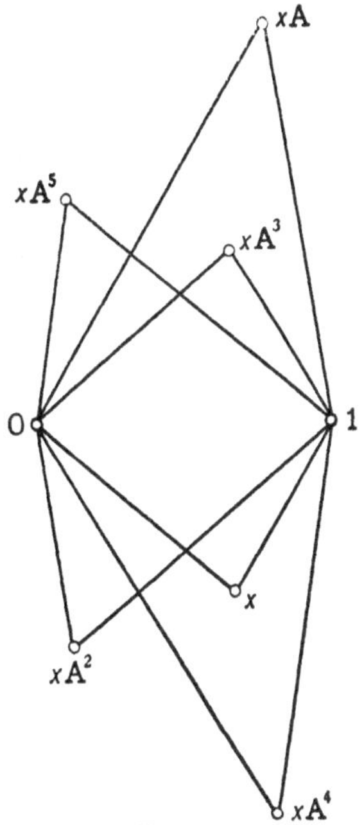

FIG. 25

§ 40. **The Cycle of Six Points** – Consider the antigraphy A which sends x_1 into x_2, x_2 into x_3, x_3 into x_1. It is of period six, so that the points of the plane fall into cycles of six. Let us consider what the cycle is which begins with ∞.

If the given points are 0, 1, ∞, then A is to send 0 into 1, 1 into ∞, ∞ into 0. It is then

$$y(1-\bar{x})=1$$

Hence y is parallel to $x-1$; that is, the circle ∞, y, 0 touches the circle 1, x, ∞.

In fig. 25 the zigzag is continued to show the set of six points: x, $x\text{A}$, $x\text{A}^2$, $x\text{A}^3$, $x\text{A}^4$, $x\text{A}^5$.

In general then

$$\begin{array}{lll} x_3,\ y,\ x_1 & \text{touches} & x_2,\ x,\ x_3 \\ x_1,\ y,\ x_2 & \text{touches} & x_3,\ x,\ x_1 \\ x_2,\ y,\ x_3 & \text{touches} & x_1,\ x,\ x_2 \end{array}$$

Taking now the general triangle x_i, or 1, 2, 3 for short, and the point x at ∞, we have a cycle of six points: ∞, $b_1, b_2, \ldots b_5$. To construct b_1 we are to have the circle 1, b_1, 3 touching the edge 2, 3, and so on (fig. 26 (*a*)). This point b_1 and the point b_5 into which ∞ passes by the reciprocal antigraphy A^5 or $\overline{321}$, are the *Brocard points* of the triangle (fig. 26 (*b*)).

Since $\overline{123}^3$ is $\bar{1}\,\bar{2}\,\bar{3}$, that is inversion in the circumcircle, b_3 is the circumcentre, b_4 is the inverse of b_1, and b_2 the inverse

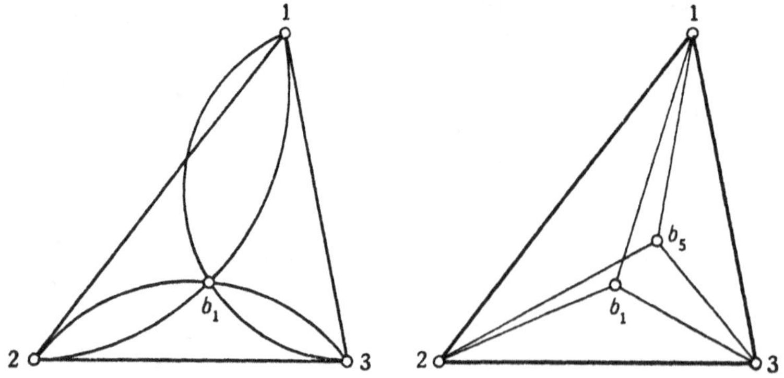

(*a*) FIG. 26 (*b*)

of b_5. These points b_2 and b_4 are simpler than the Brocard points, being the points into which ∞ is sent by the homographies $\overline{123}^2$ or (321), and $\overline{321}^2$ or (123). They might be called the Beltrami points and their join the Beltrami line, since they occur in Beltrami's memoir, *Mem. della Accad. di Bologna*, ser. 2, v. 9 (1870), where the theory of the triangle was first adequately discussed.

Exercise 2 – A is

$$(x - b_5)(\bar{y} - \bar{b}_1) = \kappa$$

Hence (fig. 26 (*b*)) deduce that

$$\angle\, b_1 23 = \angle\, b_1 31 = \angle\, b_1 12$$
$$\angle\, 23 b_5 = \angle\, 31 b_5 = \angle\, 12 b_5$$

Exercise 3 – The triangles $b_3 b_4 1$, $b_4 2 b_3$, $3 b_3 b_4$, 123 are positively similar.

Exercise 4 – The points b_3, b_4 are such that their tangents to the circumcircle are equal to their distance.

The set of 12 points, under G_{12}, which contains ∞ is

(1) the centres of the Apollonian circles—that is, the points

where the tangents to the circumcircle at x_i meet the opposite edges ;

(2) the inverses of these as to the circumcircle—that is, the mid-points of the chords $x_i j_i$ of the circumcircle, where j_i is the counter-triad ;

(3) the cycle ∞ , b_1, b_2, b_3, b_4, b_5.

Any operation of G_{12} sends the set into itself. Thus $\bar{1}\,\bar{2}\,\bar{3}$ sends the circle on which lie (23), (31), (12), b_1, b_3, b_5 into the line on which lie $\overline{23}$, $\overline{31}$, $\overline{12}$, b_2, b_4, ∞ . This circle is called the Brocard circle.

§ **41. Intrinsic Co-ordinates** — To name a point x with reference to three points x_i we may take the angles θ_1, θ_2, θ_3 which the arcs x_2, x, x_3 ; x_3, x, x_1 ; x_1, x, x_2 make with the circle x_1, x_2, x_3. These angles are to the modulus π (fig. 27).

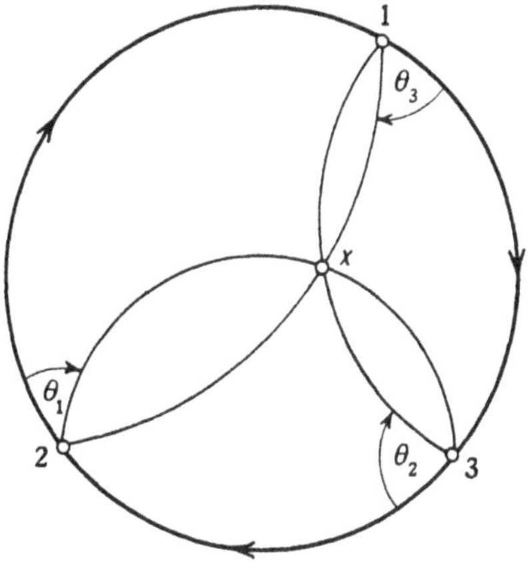

FIG. 27

Exercise 5 – The θ_i are the angles of the cross-ratios

$$\frac{x-x_3}{x-x_2}\cdot\frac{x_1-x_2}{x_1-x_3},\qquad \frac{x-x_1}{x-x_3}\cdot\frac{x_2-x_3}{x_2-x_1},\qquad \frac{x-x_2}{x-x_1}\cdot\frac{x_3-x_1}{x_3-x_2}$$

When x is ∞ , the three circles become three lines, and from this or from the exercise we have

$$\theta_1+\theta_2+\theta_3=0 \quad \text{mod } \pi$$

This is then the condition that the three circles meet on a point.

If a_i be the interior angles of the triangle

$$\theta_1=\angle\, x_2,\ x,\ x_3-a_1$$
$$\theta_2=\angle\, x_3,\ x,\ x_1-a_2$$
$$\theta_3=\angle\, x_1,\ x,\ x_2-a_3$$

In Euclidean geometry it is convenient to regard the angles θ_i as the angles of the pedal triangle of x, formed by the feet of the perpendiculars from x on the edges of the triangle. We take then

$$\theta_1 + \theta_2 + \theta_3 = \pi$$

Exercise 6 – The barycentric co-ordinates of x, in terms of θ_i, are

$$\sin^2 a_i(\cot a_i + \cot \theta_i)$$

Exercise 7 – The three points x_i, joined to x, give on the circumcircle an inverse three-point. Prove that it is positively similar to the pedal triangle of x.

We call the point named by θ_i the point θ. The points on the circumcircle other than x_i are exceptional, for here each θ_i is 0 (mod π). Also the base-points are exceptional. For x_1, $\theta_1 = 0$ and θ_2 and θ_3 are arbitrary subject to $\theta_2 + \theta_3 = 0$.

The Apollonian circles are

$$\theta_2 = \theta_3$$
$$\theta_3 = \theta_1$$
$$\theta_1 = \theta_2$$

They meet at the points $\theta_1 = \theta_2 = \theta_3$.

Since $\theta_1 + \theta_2 + \theta_3 = 0$, mod π, there are two such points

$$\theta_1 = \theta_2 = \theta_3 = \pm \pi/3$$

These are the Hessian points.

For an ordered three-point, it is convenient to call them the positive and negative Hessian points.

Since an inversion changes the sign of an angle, the inverse of θ as to the circle 1 2 3 is $-\theta_1$, $-\theta_2$, $-\theta_3$.

Exercise 8 – The point ∞ is $-a_i$. Its inverse, the circumcentre, is then a_i. Verify this.

An Apollonian inversion, say $\overline{23}$, sends θ_1, θ_2, θ_3 into θ_1, θ_3, θ_2.

Thus we have for the group G_{12} the table

Identity:	θ_1	θ_2	θ_3
$\bar{1}\ \bar{2}\ \bar{3}$	$-\theta_1$	$-\theta_2$	$-\theta_3$
$\overline{23}$	θ_1	θ_3	θ_2
$\overline{31}$	θ_3	θ_2	θ_1
$\overline{12}$	θ_2	θ_1	θ_3
(23)	$-\theta_1$	$-\theta_3$	$-\theta_2$

(31)	$-\theta_3$	$-\theta_2$	$-\theta_1$
(12)	$-\theta_2$	$-\theta_1$	$-\theta_3$
(123)	θ_2	θ_3	θ_1
(321)	θ_3	θ_1	θ_2
$\overline{123}$	$-\theta_2$	$-\theta_3$	$-\theta_1$
$\overline{321}$	$-\theta_3$	$-\theta_1$	$-\theta_2$

§ 42. **Geometric Solution of the Cubic Equation** – Let the given cubic equation be

$$a_0x^3+3a_1x^2+3a_2x+a_3=0$$

The problem is, given the numbers a_i, to construct the points.

There are two Hessian points h_i, and the cubic in terms of these is

$$(x-h_1)^3=\lambda(x-h_2)^3$$

Thus we take

$$\begin{aligned} 1-\lambda&=a_0\\ h_1-\lambda h_2&=-a_1\\ h_1^2-\lambda h_2^2&=a_2\\ h_1^3-\lambda h_2^3&=-a_3 \end{aligned}$$

whence

$$\begin{aligned} a_0a_2-a_1^2&=-\lambda(h_1-h_2)^2\\ a_0a_3-a_1a_2&=\lambda(h_1-h_2)^2(h_1+h_2)\\ a_1a_3-a_2^2&=-\lambda(h_1-h_2)^2(h_1+h_2) \end{aligned}$$

Thus the Hessian points are given by

$$(a_0a_2-a_1^2)h^2+(a_0a_3-a_1a_2)h+a_1a_3-a_2^2=0$$

This might be written directly if we assumed the homographic theory of the cubic.

We suppose this quadratic to be solved. We have also

$$3(h_1-\lambda h_2)=(1-\lambda)(x_1+x_2+x_3)$$

Thus if g be the mean point or centroid

$$g(1-\lambda)=h_1-\lambda h_2$$

or

$$\lambda=(g-h_1)/(g-h_2)$$

The solutions are then

$$(x-h_1)/(x-h_2)=\text{a cube root of }\lambda$$

Thus

$$|\ (x-h_1)/(x-h_2)\ |=|\ \lambda^{1/3}\ |$$

This gives the circumcircle of the three points x_i; and the angle h_1xh_2 is a third of the angle h_1gh_2. This gives three

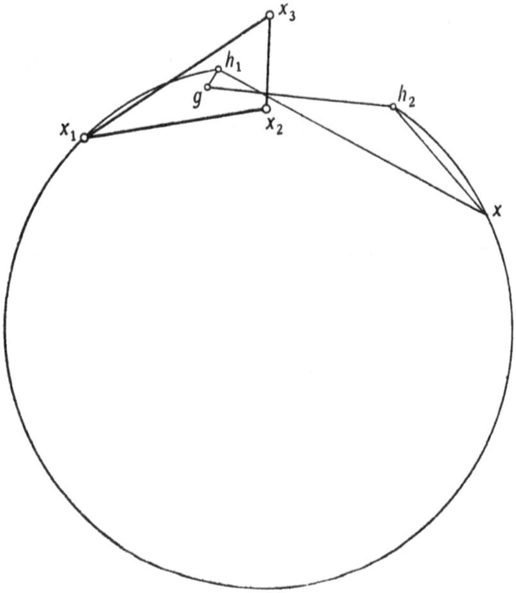

FIG. 28

arcs from h_1 to h_2, which cut the circumcircle at the points x_i (fig. 28).

§ 43. **The Groups of the Rectangle and Rhombus** – In § 26 we had the four inversions:

$$x=\bar{y}$$
$$x=-\bar{y}$$
$$x\bar{y}=1$$
$$x\bar{y}=-1$$

These form a finite group of eight operations G_8. For if we combine them we get only

$$x=y$$
$$x=-y$$
$$xy=1$$
$$xy=-1$$

Fig. 29 shows the eight fundamental regions sent into one another by the operations, one region being marked I.

These eight operations are the inversive group which sends the rectangle into itself.

Exercise 9 – For any four concyclic points, the centres of the four inversions are an orthocentric set.

In the anticyclic case, when the Jacobian of the two pairs is 0 and ∞, the parallelogram is a rhombus. The two inversions may be taken as

$$\bar{y} = x$$
$$\bar{y} = -x$$

These and their product

$$y = -x$$

form with identity the inversive group of the anticyclic pairs, G_4.

The inversive group of the rhombic four-point $\pm\rho$, $\pm\iota/\rho$ as a whole requires the inclusion of

$$x\bar{y} = \iota$$
$$x\bar{y} = -\iota$$

and

$$xy = \iota$$
$$xy = -\iota$$

Thus it is a G_8.

As permutations, the operations are

I, (34), (12), (12)(34), (1324), (1423), (13)(24), (14)(23)

The group is then the permutation group G_8.

For the group of the rhombus, the fundamental regions are as in fig. 30.

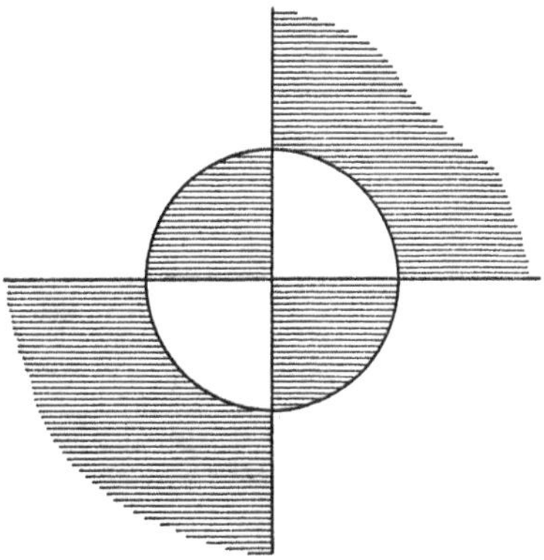

FIG. 29

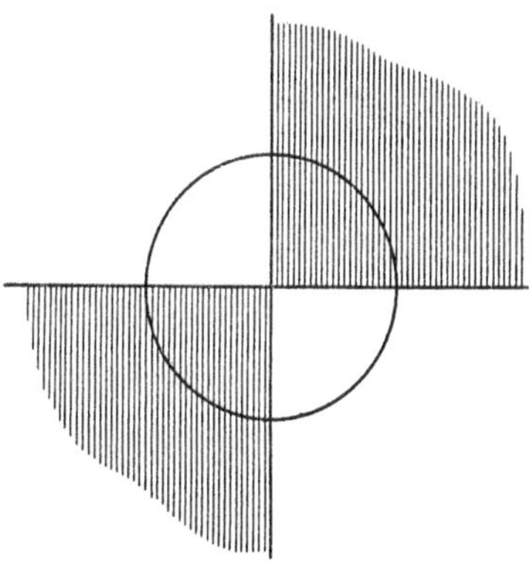

FIG. 30

§ 44. Doubly-special Four-points – A four-point may be both concyclic and anticyclic. In that case the canonical form is both a rectangle and a rhombus—that is, a square. Taking the vertices as 1, ι, -1, $-\iota$ the operations which send it into itself are:

$$
\left.
\begin{aligned}
\bar{y} &= x \\
\bar{y} &= \iota x \\
\bar{y} &= -x \\
\bar{y} &= -\iota x \\
x\bar{y} &= 1 \\
x\bar{y} &= \iota \\
x\bar{y} &= -1 \\
x\bar{y} &= -\iota
\end{aligned}
\right\}
\quad \text{and} \quad
\left\{
\begin{aligned}
y &= x \\
y &= \iota x \\
y &= -x \\
y &= -\iota x \\
xy &= 1 \\
xy &= \iota \\
xy &= -1 \\
xy &= -\iota
\end{aligned}
\right.
$$

These sixteen are the inversive group of the square, a G_{16}. If a denotes any fourth root of 1, the group is

$$y = ax, \qquad xy = a, \qquad \bar{y} = ax, \qquad x\bar{y} = a$$

A four-point can be anticyclic in more than one way. The rhombus is then such that there is a circle on two adjacent points x_1, x_2 about the remaining two, x_3, x_4.

Exercise 10 – The points x_1, x_2, x_3 are then consecutive points of a regular dodecagon.

But we shall in this case suppose one of the points themselves to be ∞. Thus we have a three-point and ∞. In the anticyclic case the three-point forms an isosceles triangle. If a triangle be doubly isosceles it is equilateral or regular. Thus our case is that of a regular three-point and ∞ (one of the Hessian points). By a suitable inversion this becomes the vertices of a regular tetrahedron, and the symmetry is manifest.

We take then the points 1, ω, ω^2, ∞. In § 39 we had the G_6 which sends the regular three-point into itself, leaving ∞ fixed. We combine with this the group G_4 of involutions which sends β, γ, ∞ into itself, leaving α fixed; where α, β, γ are 1, ω, ω^2 in any order. Writing first the homographies

$$y = ax$$
$$(x - a)(y - a) = 3a^2$$
$$(x - \beta)(y - \gamma) = 3a^2$$

we thus obtain $3+3+6$ or 12 homographies. Combining with $y=\bar{x}$, we have

$$y=a\bar{x}$$
$$(x-\alpha)(\bar{y}-\alpha)=3\alpha^2$$
$$(x-\beta)(\bar{y}-\gamma)=3\alpha^2$$

that is, 12 antigraphies. The group is thus a G_{24}. It is the symmetric group of all permutations of four things. We see this as follows. Regarding it as all reflexions and rotations of a regular tetrahedron p_i into itself, let δ_i be the distance of any other point from p_i. Under the operations the new distances δ_i' will be the old distances in a new order.

The 12 homographies are the group G_{12} of even permutations of four things, called the alternating group.

The proper canonical form for the three quadratics is the vertices of a regular octahedron. The involutions are now respectively

Q_1:	$x+y=0$,	with fixed points	$0, \infty$
Q_2:	$xy=1$,	,, ,,	± 1
Q_3:	$xy=-1$,	,, ,,	$\pm \iota$

They assign to any point x the points

$$xQ_1=-x$$
$$xQ_2=1/x$$
$$xQ_3=-1/x$$

Thus

$$xQ_1Q_2=xQ_3$$

or

$$Q_1Q_2=Q_3$$

or

$$Q_1Q_2Q_3=I$$

and similarly any product of the three is identity.

Any point x_0 is then one of a set of four, given by

$$(x-x_0)(x+x_0)(x-1/x_0)(x+1/x_0)=0$$

or

$$x^4-(x_0^2+1/x_0^2)x^2+1=0$$

This, involving only the one number (x_0^2+1/x_0^2), is a pencil of quartics.

§ 45. **The Regular Polyhedra** – In general for a finite inversive group of a plane or sphere into itself, first, all the operations must be cyclic. Second, all the extra points must be the same. For two elliptic inversions being given by chords on points a, b within the sphere, set up on the sphere a homography of the hyperbolic type, which cannot be cyclic.

The only extra points are then a single pair. Taking these as the centre and infinity, we shall have on the sphere either vertices of a regular polyhedron, or those of a regular polygon, as the fixed points of the finite inversive group. We touch here the fundamental work of Klein on the Icosahedron, where the finite groups of homographies are discussed.

To indicate how the science develops, let us mention the modular group. We take the two homographies $y = x + 1$, $yx = -1$. These send the upper half-plane into itself. Their products are of the form $y = (\alpha x + \beta)/(\gamma x + \delta)$ where α, β, γ, δ are integers such that $\alpha\delta - \beta\gamma = 1$. For a fundamental region we may take that within the verticals $x + \bar{x} = \pm 1$ and above the base-circle. This is a quadrangle with vertices $\iota\infty$, ω, ι, $-\omega^2$ and interior angles o, $\pi/_3$, π, $\pi/_3$. This group, with its subgroups, is discussed in full in Klein-Fricke, *Elliptische Modulfunktionen*. An introduction is in L. R. Ford's *Automorphic Functions* (McGraw-Hill, New York).

CHAPTER VIII

PARABOLIC, HYPERBOLIC, AND ELLIPTIC GEOMETRIES

§ 46. **Analytic Expressions for the Three Subgroups** – We select now from all inversions of a plane or sphere into itself those which leave unaltered as a whole (1) a point, (2) a circle, (3) an extra point. We thus get three subgroups of the inversive group.

Let us obtain the expressions for the antigraphies in each subgroup. We write the general antigraphy

$$\bar{y} = \frac{ax+b}{cx+d} \tag{1}$$

If we are not to disturb the point ∞ —that is, if we discard all inversions other than reflexions—then to $x=\infty$ corresponds $y=\infty$, $\bar{y}=\infty$. Hence $c=0$.

The subgroup which governs the Euclidean geometry is then formed by

$$\bar{y} = ax+b \tag{2}$$

For the hyperbolic subgroup with a fixed circle, consider first the base-line. This is fixed as a whole when to a real x corresponds a real y—that is, when (1) is

$$r = \frac{a\rho+b}{c\rho+d} = \frac{\bar{a}\rho+\bar{b}}{\bar{c}\rho+\bar{d}}$$

for more than two values of ρ. The requirement is that

$$\bar{a}=\kappa a, \qquad \bar{b}=\kappa b, \qquad \bar{c}=\kappa c, \qquad \bar{d}=\kappa d$$

so that κ is a turn. By dividing throughout by $\kappa^{1/2}$ we have

$$r = \frac{a\rho+b}{c\rho+d}$$

where a, b, c, d are now real. This is the three-parameter group of transformations of the base-line into itself. The group over the plane which leaves the base-line fixed is built from

$$\bar{y}=\frac{ax+b}{cx+d}$$

where a, b, c, d are real.

Next we fix the base-circle. Then to any turn $x=t$ corresponds a turn $\bar{y}=\tau$, so that

$$\tau=\frac{at+b}{ct+d}$$

whence

$$\frac{at+b}{ct+d}=\frac{\bar{c}+\bar{d}t}{\bar{a}+\bar{b}t}$$

If this holds for more than two values of t, then

$$a=\kappa\bar{d}, \qquad b=\kappa\bar{c}, \qquad c=\kappa\bar{b}, \qquad d=\kappa\bar{a}$$

whence $\kappa\bar{\kappa}=1$ and κ is a turn. Thus τ is

$$\frac{at/\sqrt{\kappa}+b/\sqrt{\kappa}}{\sqrt{\kappa}\bar{b}t+\sqrt{\kappa}\bar{a}}$$

which we write

$$\tau=\frac{at+\beta}{\bar{\beta}t+\bar{a}}$$

This is the three-parameter group of transformations of the base-circle into itself; and the group over the plane which leaves this circle fixed is built from

$$\bar{y}=\frac{ax+\beta}{\bar{\beta}x+\bar{a}} \tag{3}$$

The third case is that of a fixed elliptic inversion; we take this to be

$$x\bar{x}_1+1=0$$

Then

$$\bar{y}=\frac{ax+b}{cx+d}$$

and

$$\bar{y}_1=\frac{ax_1+b}{cx_1+d}$$

where both

$$y\bar{y}_1+1=0$$
$$x\bar{x}_1+1=0$$

That is,

$$\frac{ax+b}{cx+d} = -1/y_1 = -\frac{\bar{c}\bar{x}_1+\bar{d}}{\bar{a}\bar{x}_1+\bar{b}}$$

$$= \frac{-\bar{c}+\bar{d}x}{\bar{a}-\bar{b}x}$$

for all values of x. Hence

$$a = \kappa\bar{d}, \qquad b = -\kappa\bar{c}, \qquad c = -\kappa\bar{b}, \qquad d = \kappa\bar{a}$$

Here κ is again a turn, and

$$\bar{y} = \frac{ax/\sqrt{\kappa}+b/\sqrt{\kappa}}{-\bar{b}\sqrt{\kappa}x+\bar{a}\sqrt{\kappa}}$$

which we write

(4) $$\bar{y} = \frac{ax+\beta}{-\bar{\beta}x+\bar{a}}$$

This is then the general operation of the elliptic geometry.

§ 47. **Infinity** – The three subgroups give three types of two-way geometry: (1) the parabolic with a fixed point Ω, (2) the hyperbolic with a circle Ω fixed as a whole, (3) the elliptic with a fixed extra pair Ω.

We are in each case restricted to the inversions which leave Ω fixed and their products. An even number of such inversions gives a motion. A card lying on the two-way can be freely moved, but it cannot by a finite number of motions reach Ω. What is fixed is inaccessible, or infinity.

Consider the first type. If one is drawing figures on a sheet of paper, one can isolate any point Ω of the paper. The lines are now circles on Ω. But if one is considering his relation to a plane on which he is, he regards Ω as what he cannot reach. This is Euclidean geometry. The point ∞ is attached to the number ∞. This is an egocentric view, but it is very useful.

Under an inversion, from an outside point, the plane is a sphere with the point ∞ marked on it. What seem to a man on the sphere to be equal steps in a given direction are given by secants which touch an ellipse. The background is here a Euclidean space.

In the elliptic case (3) we invert with regard to one of the points Ω. We have then geometry on a sphere. The fixed inversion is the interchange of antipodal points. If we choose to allow this, we have ordinary spherical geometry. The lines of this geometry are great circles. The distance of two points x_1, x_2 is the angle made by the radii to x_1 and x_2. It is thus an elliptic distance.

If we do not allow the interchange of antipodal points, we have the strict elliptic geometry. We may take as a point a diameter of the sphere (not directed) and as a line a diametral plane. In other words, we have lines on a point as points and planes on that point as lines, in a Euclidean space.

We consider especially the hyperbolic type (2). Infinity is now a circle Ω. We consider the plane of this circle. In this plane are ∞^2 circles l normal or orthogonal to Ω. These are the fixed circles of the hyperbolic inversions which send Ω into itself.

Two of the circles, l_1 and l_2, normal to Ω may intersect, in which case the intersections are inverse as to Ω; or they may touch, in which case the point of contact is on Ω, or they may not intersect, in which case there is one circle l_3 normal to both l_1 and l_2 and to Ω.

Take now the egocentric view of the man on a plane, for whom the accessible region is bounded by the circle Ω. The outside region has only theoretic interest. There are on two points x_1 and x_2 a circle l which is to him a line. He calls it a line. On it are two points at ∞—that is, on Ω. Let these be j and j'.

He observes that, if two lines do not meet, they will have either an infinity of common normals or just one. He calls them in the former case parallel, and sees that through a point are two parallels to a given line, one to j and one to j'. His geometry is that of Euclid except in the fact of the two parallels, or in the fact that there is no rectangle.

From the notion of reflexion in a line l (Hjelmslev's axiom) he can build up rigid motions, but he has no translations. Not having parallelograms, he has no parallelogram law. His algebra does not readily apply to his geometry. His analytic geometry is hyperbolic trigonometry.

§ 48. **Distance** – The notion of angle requires no alteration when infinity is a circle. For the distance of two points x_1

and x_2 we have now on the joining arc two end-points on Ω—the points at ∞. Let these be j and j'. This gives to x_1 and x_2 a hyperbolic distance. Let us express it by means of x_1, x_2 and their inverses $x_1\Omega$ and $x_2\Omega$.

An inversion with regard to j or j' sends Ω into a line, and x_1 and x_2 into points on a perpendicular line. Thus x_1, x_2, $x_1\Omega$, $x_2\Omega$ may be taken as 1, p, -1, $-p$. The cross-ratio $(1, -p \mid p, -1)$ is

$$-\frac{(1-p)^2}{4p}$$

The stretch which sends 1 into p being e^{λ}, we have

$$(1, -p \mid p, -1) = -\sinh^2 \lambda/2$$

And therefore in general the hyperbolic distance from x_1 to x_2 is given by

$$\sinh^2 \lambda/2 = -(x_1, x_2\Omega \mid x_2, x_1\Omega)$$

When Ω is the axis of reals, this is

$$\sinh^2 \lambda/2 = -(x_1, \bar{x}_2 \mid x_2, \bar{x}_1)$$
$$= \delta_{12}{}^2/4\eta_1\eta_2$$

where δ_{12} is the parabolic distance of the points and η_i the parabolic distance to the axis. For a small displacement this gives

$$d\lambda = ds/\eta$$

which is taken as fundamental by Poincaré.

The distance of two points x_i with reference to an inversion Ω is in general given by

$$\sin^2 v/2 = (x_1, x_2\Omega \mid x_2, x_1\Omega)$$

where v is θ or $-\iota\lambda$, according as the inversion is elliptic or hyperbolic.

If the inversion is $x\bar{y} = \kappa$ the formula is

$$\sin^2 v/2 = (x_1, \kappa/\bar{x}_2 \mid x_2, \kappa/\bar{x}_1)$$
$$= -\kappa\frac{(x_1 - x_2)(\bar{x}_1 - \bar{x}_2)}{(x_1\bar{x}_1 - \kappa)(x_2\bar{x}_2 - \kappa)}$$

or if p_{ij} is the power or squared distance of x_i and x_j,

$$\sin^2 v/2 = -\kappa\frac{p_{12}}{(p_{01} - \kappa)(p_{02} - \kappa)}$$

Exercise 1 – At 0 erect a normal 0c to the plane. Let the arcs *ca*, *cb* be normal to the plane at *a* and *b*. Show that the angle between the arcs is given by

$$\sin \theta/2 = | c(a-b)/ab |$$

§ 49. **Curvature** – Another way of combining the two cases very valuable for general geometry is to write $\kappa = -\kappa_0^2$, $v = u/\kappa_0$, thus introducing a unit of distance. The above formula becomes

$$\sin^2 u/2\kappa_0 = \kappa_0^2 p_{12}/(p_{01} + \kappa_0^2)(p_{02} + \kappa_0^2)$$

Thus for geometry on a sphere of radius κ_0 the fixed inversion is $\rho\rho' = -\kappa_0^2$ (sending each point of the sphere into its antipodal point). The arc u is $\kappa_0\theta$, p_{0i} is κ_0^2, and the formula gives, as it should,

$$\sin^2 \theta/2 = p_{12}/4\kappa_0^2$$

This applies equally to a circle, or to any elliptic spread, the radius being κ_0. In the case of the circle, the *curvature* is $1/\kappa_0$; in the case of the sphere the (Gaussian) curvature is $1/\kappa_0^2$. In general for an n-way it would be $1/\kappa_0^n$. Thus an elliptic spread is of constant positive curvature, the curvature being estimated parabolically from the extra points, and depending solely on the constant κ^2 or $-\kappa$. In the case of a positive κ, for which κ_0 is a pure imaginary, we have constant negative curvature, only for $2n$-way spreads.

The constant real κ of an inversion $\rho\rho' = \kappa$ is then the *constant of curvature.*

The hyperbolic and elliptic geometries differ only in the sign of κ. Thus theorems proved for the one are true for the other, unless the sign of κ enters the proof. As an instance where theorems are not coextensive, but are contradictory, we have for negative κ the theorem of spherical geometry that the sum of the angles of a triangle is $> \pi$. For hyperbolic geometry, the sum $< \pi$.

§ 50. **Motions** – It is a fundamental idea that a spread with infinity Ω can be discussed directly from Ω; that the proper reference-scheme is Ω itself. This we regard as the essence of the non-Euclidean geometries.

It is obscured when Ω is a mere point, so that Euclidean geometry is here an inconvenient special case. Let us illustrate what is meant by the theory of circles on a plane.

We take in the plane a circle Ω. Let it be the axis of reals. Let H be a homography which sends Ω into itself. This is called a *motion* in the hyperbolic half-plane. The motion may be hyperbolic with fixed points on Ω, elliptic with fixed points inverse as to Ω, or parabolic with a fixed double point on Ω. It assigns to any point ρ of Ω a point ρH. We represent the ordered pair ρ, ρH by the arc in the half-plane (here a directed semicircle) with these ends, orthogonal to Ω.

It is clear in the case $y = \kappa x$, where κ is a real, that all the

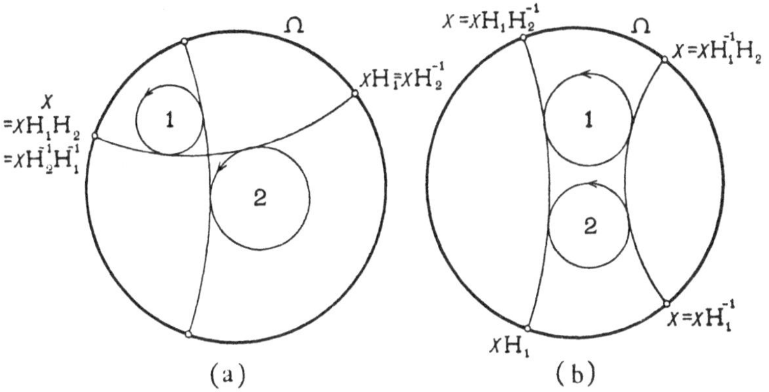

FIG. 31

arcs for a given H touch a circle. And generally, if the arc ρ_1, ρ_2 touches a circle (x, r) positively, then

$$[x - (\rho_1 + \rho_2)/2][\bar{x} - (\rho_1 + \rho_2)/2] - [r_1 - (\rho_1 - \rho_2)/2]^2$$

or if ξ, η be the rectangular co-ordinates of the centre x,

$$(1) \qquad \rho_1\rho_2 - (\xi - r)\rho_1 - (\xi + r)\rho_2 + \xi^2 + \eta^2 - r^2 = 0$$

This is the general motion or homography with fixed Ω. The arcs of H then all touch a circle, and the circle is directed since the arcs are.

Exercise 2 – A change in the sign of r gives the reciprocal motion.

A directed circle then represents the motion H, and we call the circle also H. The motion is hyperbolic, parabolic, or elliptic as the circle cuts, touches, or does not meet Ω.

For two motions H_1, H_2 we have two directed circles. When negative common tangent arcs exist (fig. 31 (a)) they have as their end-points the fixed points of H_1H_2 and those of H_2H_1. When positive common tangent arcs exist (fig. 31 (*b*)), they give the common pairs of H_1, H_2—that is, the fixed points of $H_1H_2^{-1}$ (or of $H_2H_1^{-1}$) and of $H_2H_1^{-1}$ (or of $H_1H_2^{-1}$).

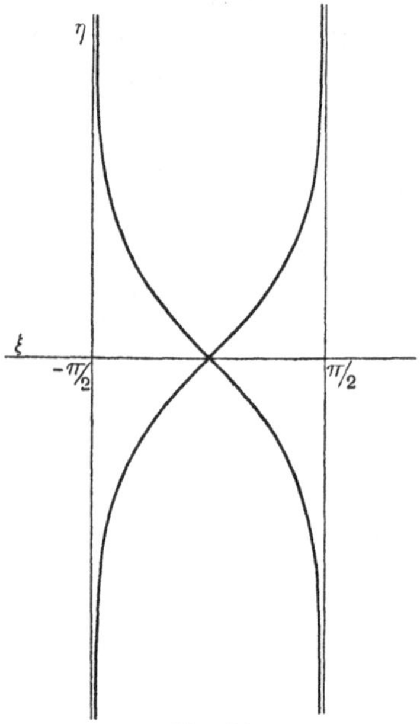

FIG. 32

A motion, elliptic or hyperbolic, carries any point through a certain distance, elliptic or hyperbolic. For the motion

$$H \equiv \alpha xy + \beta x + \gamma y + \delta$$

where α, β, γ, δ are real, the distance is given by

$$\cos v = g_1/2\sqrt{g_2}$$

where $g_1 = \beta - \gamma$, $g_2 = \alpha\delta - \beta\gamma$ (§ 34).

When $g_1^2 < 4g_2$, v is an angle θ, the elliptic distance or angle of the motion. When $g_1^2 > 4g_2$, v is $-\iota\lambda$, where λ is the hyperbolic distance. Let us compare this v with the vector v' of the two circles Ω and H. The invariants of (1) are

$$g_1 = 2r, \qquad g_2 = \eta^2$$

and v' is given (§ 25) by

$$\cos v' = \eta/r$$
$$= 2\sqrt{g_2}/g_1$$

Hence, as in § 25,

$$\cos v \cos v' = 1 \tag{2}$$

That is, the cosines are reciprocal. When v is a real, say ξ, v' is a pure imaginary, $\iota\eta$, the relation of ξ and η being indicated in fig. 32.

§ 51. **The Vector of Two Homographies** – Two homographies have in addition to their individual invariants the invariant

$$2g_{11} = \alpha\delta' + \alpha'\delta - \beta\gamma' - \beta'\gamma$$

From two equations such as § 50 (1) we have then

$$2g_{11} = (\xi_1 - \xi_2)^2 + \eta_1{}^2 + \eta_2{}^2 - (r_1 - r_2)^2$$

(1)
$$= 2\eta_1\eta_2 + p_{12} - (r_1 - r_2)^2$$

where p_{12} is the squared Euclidean distance of the Euclidean centres.

If now the circles touch positively

$$p_{12} = (r_1 - r_2)^2$$

so that

(2)
$$g_{11}{}^2 = g_{20}g_{02}$$

This is the condition for positive contact.

Exercise 3 – The condition for negative contact is

$$(g_{11} - g_{10}g_{01}/2)^2 = g_{20}g_{02}$$

When there is an arc which touches both circles positively, let ∞ be a point where the arc meets Ω.   Then

$$\xi_1 - \xi_2 = r_1 - r_2$$

whence (1) becomes

$$2(g_{11} - \eta_1\eta_2) = (\eta_1 - \eta_2)^2$$

By § 48 this is

$$4\eta_1\eta_2 \sinh^2 \lambda_{12}/2$$

where λ_{12} is the length of the common tangent.   Thus the hyperbolic length of the common tangent is given by

(3)
$$\cosh \lambda_{12} = g_{11}/\sqrt{g_{20}}\sqrt{g_{02}}$$

These are two such tangents; the lengths are equal.

In any case the vector v attached to $H_1H_2{}^{-1}$ is given by

$$\cos v = g_{11}/\sqrt{g_{20}}\sqrt{g_{02}}$$

For let H_1 be $\alpha xy + \beta x + \gamma y + \delta$, and H_2 be $\kappa x - y$.   Then

$$H_2{}^{-1} \text{ is } x - \kappa y \quad \text{or} \quad y - \kappa z$$
$$H_1H_2{}^{-1} \text{ is } \kappa\alpha xz + \beta x + \kappa\gamma z + \delta$$

with invariants

$$g_1{}' = \beta - \kappa\gamma, \qquad g_2{}' = \kappa(\alpha\delta - \beta\gamma)$$

And for H_1 and H_2

$$g_{20} = \alpha\delta - \beta\gamma, \qquad g_{02} = \kappa, \qquad 2g_{11} = \beta - \kappa\gamma$$

so that

$$g_{11}{}^2/g_{20}g_{02} = g_1{}'^2/4g_2{}'$$
$$= \cos^2 v$$

We have attached to the homography H which sends Ω into itself a directed circle. For the pencil $H_1 + \lambda H_2$ we have a pencil of directed circles. These are all the circles which touch positively two given circles, namely, the two circles common to H_1 and H_2. The equations $H_1 = 0$, $H_2 = 0$ give two values of x, say x_1, x_2; and the corresponding values of y, say y_1, y_2. The common circles are x_1, y_1 and x_2, y_2.

For three circles which have an orthogonal circle Ω there is a convenient canonical form. We have on Ω three ordered pairs. These lie in a homography H. Write this $y = \kappa^2 x$. The ordered pairs are now

$$\alpha/\kappa,\ \alpha\kappa; \qquad \beta/\kappa,\ \beta\kappa; \qquad \gamma/\kappa,\ \gamma\kappa$$

The circle H is one of the two which touch the three directed circles positively. The standard case is when κ is a turn.

It is an easy matter, by considering the Jacobians of these pairs, to develop what corresponds to spherical trigonometry. But this will appear as a special case in the next chapter (§ 56).

CHAPTER IX

THE CELESTIAL SPHERE

§ 52. **Geometry within a Sphere** – Let us regard the space around us as bounded by a sphere Ω. The points of Ω are named by numbers x. To two points x and x' we attach that arc of the circle normal to Ω at x and x' which lies in our space. This arc we call * a line of our space; its end points are x and x'. We regard it as directed from x to x'. But if x and x' are only given collectively by a quadratic q,

$$q = \alpha x^2 + 2\beta x + \gamma$$

then the line is not directed.

Two directed lines give a vector $v \equiv \theta - \iota\lambda$, by the formula (§ 29)

$$\cos(\theta - \iota\lambda) = \frac{(x_1 + x_1')(x_2 + x_2') - 2x_1x_1' - 2x_2x_2'}{(x_1 - x_1')(x_2 - x_2')}$$

For intersecting lines $\lambda = 0$. Two intersecting lines lie in a plane of the space (in the Euclidean space a sphere normal to Ω). But in a plane two lines may not intersect. In that case $\theta = 0$ or π. And in a plane two lines may be parallel—that is, may meet on Ω (touching there). In that case $\lambda = 0$ and $\theta = 0$ or π.

Thus in the formula let $x_2 = x_1$. Then $\cos(\theta - \iota\lambda) = 1$ and $\lambda = 0$, $\theta = 0$ or π.

For perpendicular lines $\theta = \pi/2$. For lines which are normal—that is, both intersecting and perpendicular—$\lambda = 0$ and $\theta = \pi/2$.

* There are here two languages, one for the Euclidean space and one for the region within the sphere. In this latter we have geodesic lines, geodesic planes, geodesic cylinders, which are called in the former arcs orthogonal to the sphere, spheres orthogonal to the sphere, Dupin cyclides. Until the end of § 63 we use the intrinsic or geodesic naming; the word geodesic is to be understood.

The proof that the shortest path is along the arc of a circle, when distance is defined by Poincaré's formula $\dot{x}_1^2 + \dot{x}_2^2 + \dot{x}_3^2 = x_1^2$ belongs to the Calculus of Variations. The essential fact is that $\dot{x}_2 = \kappa_2 x_1^2$ and $\dot{x}_3 = \kappa_3 x_1^2$, for a geodesic.

Thus two quadratics

$$q_i = a_i x^2 + 2\beta_i x + \gamma_i$$

give normal lines when the bilinear invariant or power

$$q_{12} = a_1\gamma_2 - 2\beta_1\beta_2 + \gamma_1 a_2$$

vanishes.

Two lines have one common normal, given by the Jacobian of the two quadratics,

$$j_{12} = \begin{vmatrix} 1 & -x & x^2 \\ a_1 & \beta_1 & \gamma_1 \\ a_2 & \beta_2 & \gamma_2 \end{vmatrix}$$

In passing from the line x_1, x_1' to the line x_2, x_2' along the common normal j, j', θ is the angle made by the plane j, j', x_2, x_2' with the plane j, j', x_1, x_1'. And λ is the distance gone—the hyperbolic distance reckoned from j and j'.

An airplane with regard to the celestial sphere is defined by two normal lines, say j, j' from stern to bow, and x_1, x_1' from port wing to starboard wing. For these lines $\cos v = 0$—that is, $\lambda = 0$ and $\theta = \pi/2$. The two harmonic pairs of numbers, when ordered, define the position. They give six parameters, but these are not wanted. As the airplane travels along the line j, j' the angle θ turned through sidewise and the hyperbolic distance λ gone are given by (1).

§ 53. **The Directed Cylinder** – Taking now Ω as a plane, we consider on it the general homography H. We attach to the points x and xH an arc (of a circle) normal to the plane.

Let H be in the form

$$\kappa^2 x = y$$

When κ is a real, evidently all the arcs touch a certain right cone with vertex 0. When κ is a turn, the arcs still touch this cone. Hence for any κ the arcs will touch a right cone, the common tangent making a fixed angle θ with the generator.

If we adopt the language of § 52, then when Ω is a sphere what represents a homography H on the sphere is a cylinder whose axis is the line joining the fixed points; and the cylinder is marked with an arrowhead to denote the angle θ. Since H is determined by three ordered pairs x_i, x_i', the directed cylinder is determined by three directed lines, which it touches at the same angle θ. We may call it the inscribed cylinder, since it is the natural generalisation of the inscribed circle.

This hyperbolic cylinder is a Dupin cyclide, the inverse of a right cone, or the inverse of the spindle obtained by rotating an arc about its chord. In the elliptic three-way it would be the inverse of an anchor ring obtained by rotating a circle about any line in its plane.

The axis of the cylinder makes the same vector $\theta - \iota\lambda$ ($=r$, say) with each of the given lines. The pairs 0, ∞ and x, $\kappa^2 x$ give

$$\cos r = (\kappa^2 + 1)/(\kappa^2 - 1)$$

or

$$\cot r/2 = \pm\iota\kappa$$
$$2 \cot r = \pm\iota(\kappa + 1/\kappa)$$

In terms of the invariants $g_1 = \kappa^2 + 1$, $g_2 = \kappa^2$ we have then

(1) $$4 \cot^2 r = -g_1{}^2/g_2$$

Exercise 1 – The cylinder where it meets Ω has a conical point. If the angle at this point be $2a$, then

$$\lambda = \log \tan (\pi/4 + a/2)$$

so that

$$\cosh \lambda \cos a = 1$$

Exercise 2 – A right cone has the angle $2a$, vertex 0. It has bitangent circles parallel to the axis. Let such a circle make at a point of contact an angle θ with the generator. And let the ratio of the greatest and least vectors from 0 to the circle be $\varrho e^{\iota\phi}$. Then

$$\varrho = (1 + \cos a \cos \theta)/(1 - \cos a \cos \theta)$$

and

$$\tan \varphi/2 = \sin \theta \cot a$$

§ 54. **The Determinant of Powers** – The theory of the lines of a hyperbolic space is the theory of quadratics. If a line q_1 is to be directed we must select a square root of its discriminant q_{11}.

Conversely a system of quadratics is properly represented by the line-geometry of a hyperbolic space—a geometry in which a line, instead of a point, is the foundation.

For four points of Ω we have three pairs of points, and thus three pairs of lines; and three common normals. The theory of the four-point shows that the three common normals meet at a point, at right angles.

We take as fundamental the fact that four quadratics q_i must be linearly related—that is, that κ_i can be so determined that

$$\kappa_1 q_1 + \kappa_2 q_2 + \kappa_3 q_3 + \kappa_4 q_4 \equiv 0 \tag{1}$$

Explicitly the identity is

$$\begin{vmatrix} q_1 & q_2 & q_3 & q_4 \\ \alpha_1 & \alpha_2 & \alpha_3 & \alpha_4 \\ \beta_1 & \beta_2 & \beta_3 & \beta_4 \\ \gamma_1 & \gamma_2 & \gamma_3 & \gamma_4 \end{vmatrix} = 0$$

Denoting the bilinear invariant of q_i and q_j by q_{ij}, then we have

$$\kappa_1 q_{i1} + \kappa_2 q_{i2} + \kappa_3 q_{i3} + \kappa_4 q_{i4} = 0$$

so that the determinant

$$| q_{ij} | \equiv | q_{i1}\, q_{i2}\, q_{i3}\, q_{i4} | = 0$$

Now let v_{ij} be the vector of the two quadratics q_i, q_j. Then

$$\cos v_{ij} = q_{ij} / \sqrt{q_{ii}} \sqrt{q_{jj}}$$

Hence, dividing the ith row of the determinant by a selected root of q_{ii} and the jth column by a selected root of q_{jj}, we have

$$| \cos v_{ij} | = 0 \tag{2}$$

where v_{ij} is the vector from the directed line q_i to the directed line q_j.

There is a useful extension of (2). For if we take four other quadratics $q_{i'}$, the bilinear invariants with (1) give

$$\kappa_1 q_{1i'} + \kappa_2 q_{2i'} + \kappa_3 q_{3i'} + \kappa_4 q_{4i'} = 0$$

whence for two sets of four quadratics

$$| q_{ii'} | = | q_{i1'}\, q_{i2'}\, q_{i3'}\, q_{i4'} | = 0$$

and

$$| \cos v_{ii'} | = 0 \tag{3}$$

We denote this determinant by D_4, the context showing whether the general or the symmetric case is meant. The determinant of three rows, whose elements are $\cos v_{ii'}$, where now $i=1, 2, 3$, $i'=1, 2, 3$ will be similarly denoted by D_3.

§ 55. **The Radius of the Cylinder** – The developed form of this determinant D_4 in (2) above is, writing c_{ij} for $\cos v_{ij}$,

$$(1) \quad 1-\sum^{6} c_{12}{}^2+2\sum^{4} c_{23}c_{31}c_{12}+\sum^{3} c_{12}{}^2c_{34}{}^2-2\sum^{3} c_{12}c_{23}c_{34}c_{41}$$

Let us apply this to determine the radius of the inscribed cylinder of three lines q_1, q_2, q_3.

We have

$$c_{14}=c_{24}=c_{34}=\cos r$$

where r is the vector radius. Hence

$$\cos^2 r[3-\Sigma c_{12}{}^2+2\Sigma c_{12}c_{13}-2\Sigma c_{23}]=1-\Sigma c_{12}{}^2+2c_{23}c_{31}c_{12}$$
$$=D_3$$

whence

$$D_3 \tan^2 r=2(1-\cos v_{23})(1-\cos v_{31})(1-\cos v_{12})$$

We have, expressed in factors, if $2s=v_{23}+v_{31}+v_{12}$,

$$D_3=4 \sin s \,.\, \sin (s-v_{23}) \sin (s-v_{31}) \sin (s-v_{12})$$

and

$$3-\Sigma c_{12}{}^2+2\Sigma c_{12}c_{13}-2\Sigma c_{23}=8\Sigma \sin^2 v_{31}/2 \sin^2 v_{12}/2-4\Sigma \sin^4 v_{23}/2$$
$$=-4\overset{4}{\Pi}(\sin v_{23}/2 \pm \sin v_{31}/2 \pm \sin v_{12}/2)$$

so that

$$\cos^2 r/\sin s \,.\, \sin (s-v_{23}) \sin (s-v_{31}) \sin (s-v_{12})$$
$$=\sin^2 r/4 \sin^2 v_{23}/2 \sin^2 v_{31}/2 \sin^2 v_{12}/2$$
$$(2) \qquad =-1/\Pi(\sin v_{23}/2 \pm \sin v_{31}/2 \pm \sin v_{12}/2)$$

Comparing with § 53 (1),

$$\cos^2 r/g_1{}^2=-\sin^2 r/4g_2=1/(g_1{}^2-4g_2)$$

we have the invariants of a homography in terms of the vectors of any three pairs in it. We shall take

$$(3) \quad g_1{}^2=\tfrac{1}{4}D_3=\sin s \,.\, \sin (s-v_{23}) \sin (s-v_{31}) \sin (s-v_{12})$$

These formulæ should be associated with the meaning of the vanishing of an invariant. When $g_1=0$, the pairs are in a polarity (or the lines have a common normal). We may take then

$$x_i+x_i'=0$$

Then from

$$\cos v_{23} = (x_2{}^2 + x_3{}^2)/2x_2x_3$$
$$x_2/x_3 = e^{\pm \iota v_{23}}$$

and

$$v_{23} \pm v_{31} \pm v_{12} = 0$$

When $g_2 = 0$, the homography is singular, and

$$(x_2 - x_3)(x_3 - x_1)(x_1 - x_2)(x_2' - x_3')(x_3' - x_1')(x_1' - x_2') = 0$$

But

$$\sin^2 v_{23}/2 = (x_2 - x_3)(x_2' - x_3')/(x_2 - x_2')(x_3 - x_3')$$

so that

$$\sin^2 v_{23}/2 \sin^2 v_{31}/2 \sin^2 v_{12}/2 = 0$$

When the pairs are in a translation we may take

$$x_i' = x_i + \mu$$

We have then

$$\sin^2 v_{23}/2 = (x_2 - x_3)^2/\mu^2$$

so that

$$\sin v_{23}/2 \pm \sin v_{31}/2 \pm \sin v_{12}/2 = 0$$

Exercise 3 – In terms of r, the three cases are where

$$r = \pi/2, \qquad 0, \qquad \infty.$$

Exercise 4 – Using the identity

$$\sin(\alpha + \beta)\sin(\alpha - \beta) = \sin^2\alpha - \sin^2\beta$$

prove directly the identity

$$\sin(\alpha+\beta+\gamma)\sin(\alpha-\beta-\gamma)\sin(\beta-\gamma-\alpha)\sin(\gamma-\alpha-\beta)$$
$$-(\sin\alpha+\sin\beta+\sin\gamma)(\sin\alpha-\sin\beta-\sin\gamma)(\sin\beta-\sin\gamma-\sin\alpha)(\sin\gamma-\sin\alpha-\sin\beta)$$
$$= 4\sin^2\alpha\sin^2\beta\sin^2\gamma$$

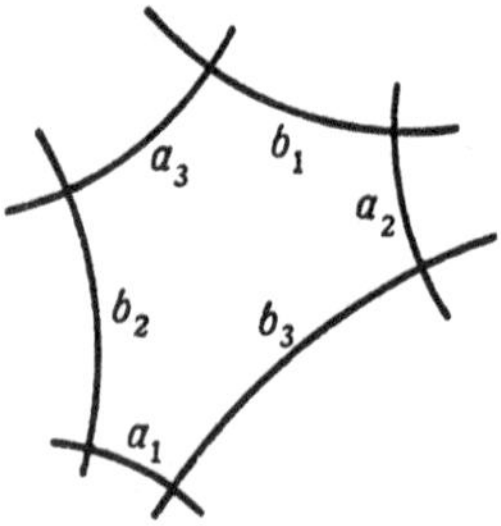

FIG. 33

§ 56. The Rectangular Hexagon – Three lines A_1, A_2, A_3 have, taken in pairs, three common normals B_1, B_2, B_3. Thus we have a rectangular hexagon (fig. 33). Let the vector edges be successively $a_1, b_3, a_2, b_1, a_3, b_2$. Apply to B_1, B_2, B_3, A_1 the formula § 54 (2).

Then

$$\begin{vmatrix} 1 & \cos a_3 & \cos a_2 & \cos c_1 \\ \cos a_3 & 1 & \cos a_1 & 0 \\ \cos a_2 & \cos a_1 & 1 & 0 \\ \cos c_1 & 0 & 0 & 1 \end{vmatrix} = 0$$

where c_1 is the vector between A_1 and B_1. Therefore

$$\cos^2 c_1 \sin^2 a_1 = \begin{vmatrix} 1 & \cos a_3 & \cos a_2 \\ \cos a_3 & 1 & \cos a_1 \\ \cos a_2 & \cos a_1 & 1 \end{vmatrix}$$

$$= \cos^2 c_2 \sin^2 a_2 = \cos^2 c_3 \sin^2 a_3$$

and equally

$$\cos^2 c_1 \sin^2 b_1 = \cos^2 c_2 \sin^2 b_2 = \cos^2 c_3 \sin^2 b_3$$

$$= \begin{vmatrix} 1 & \cos b_3 & \cos b_2 \\ \cos b_3 & 1 & \cos b_1 \\ \cos b_2 & \cos b_1 & 1 \end{vmatrix}$$

To take the square roots we must notice the formula § 55 (3),

$$\begin{vmatrix} 1 & \cos b_3 & \cos b_2 \\ \cos b_3 & 1 & \cos b_1 \\ \cos b_2 & \cos b_1 & 1 \end{vmatrix} = 4g_1^2$$

For a selected ordering we have

$$\cos c_1 \sin b_1 = \cos c_2 \sin b_2 = \cos c_3 \sin b_3 = 2g_1$$
$$\cos c_1 \sin a_1 = \cos c_2 \sin a_2 = \cos c_3 \sin a_3 = 2g_1'$$

Hence

$$\sin a_1/\sin b_1 = \sin a_2/\sin b_2 = \sin a_3/\sin b_3 = g_1'/g_1$$

This is the *law of sines.*

Let us now apply the extended formula § 54 (3) to the two sets $B_1B_2B_3A_2$ and $B_1B_2B_3A_3$. It becomes

$$\begin{vmatrix} 1 & \cos a_3 & \cos a_2 & 0 \\ \cos a_3 & 1 & \cos a_1 & 0 \\ \cos a_2 & \cos a_1 & 1 & \cos c_3 \\ 0 & \cos c_2 & 0 & \cos b_1 \end{vmatrix} = 0$$

or

$$\cos c_2 \cos c_3(\cos a_1 - \cos a_2 \cos a_3) = \cos b_1 \begin{vmatrix} 1 & \cos a_3 & \cos a_2 \\ \cos a_3 & 1 & \cos a_1 \\ \cos a_2 & \cos a_1 & 1 \end{vmatrix}$$

$$= \cos b_1 \cos c_2 \sin a_2 \cos c_3 \sin a_3$$

Hence

$$\cos a_1 - \cos a_2 \cos a_3 = \sin a_2 \sin a_3 \cos b_1$$

This is the *law of cosines.*

These two laws are the fundamental formulæ of spherical trigonometry. This special case arises when the three lines

A_1, A_2, A_3 meet at a point 0. Then B_1, B_2, B_3 are the respective normals at 0 to A_2, A_3, etc. The angles made by these normals are the angles made by the planes on A_2, A_3; A_3, A_1; A_1, A_2. The vectors are here all real—that is, they are angles.

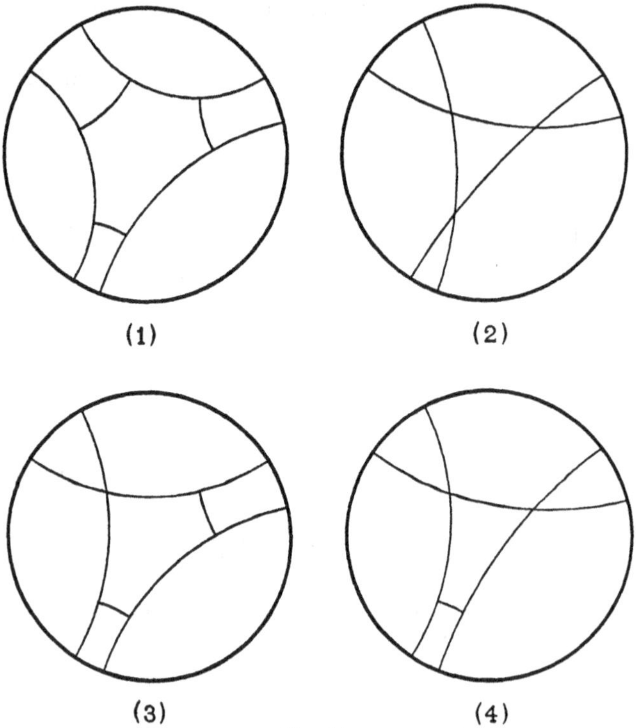

(1) (2) (3) (4)

Fig. 34

The vectors a_i are the angles and the vectors b_i the sides of a spherical triangle; or conversely.

The consequences of the two laws are then so far as analysis is concerned precisely those given in works on spherical trigonometry, and we may refer to such works.*

In the case when the three quadratics A_i give points on a circle, referred to in § 51, we take for Ω that sphere of which the circle is a great circle. In fig. 34 (1) we have

$$b_i = -\iota m_i, \qquad a_i = -\iota l_i$$

* *A Treatise on Spherical Trigonometry*, by J. Casey (Longmans Green, London), is convenient.

so that the formulæ are

$$\sinh l_1/\sinh m_1 = \sinh l_2/\sinh m_2 = \sinh l_3/\sinh m_3$$
$$\cosh l_1 - \cosh l_2 \cosh l_3 = -\sinh l_2 \sinh l_3 \cosh m_1$$

In fig. 34 (2) we have

$$b_i = \phi_i, \qquad a_i = -\iota l_i$$

In fig. 34 (3) we have

$$b_1 = \phi_1, \qquad b_2 = -\iota m_2, \qquad b_3 = -\iota m_3, \qquad a_i = -\iota l_i$$

In fig. 34 (4) we have

$$b_1 = -\iota m_1, \qquad b_2 = \phi_2, \qquad b_3 = \phi_3, \qquad a_i = -\iota l_i$$

§ 57. **The Configuration of Ten Lines** – Let us find the condition that the common normals J_{12} and J_{34} of two pairs of lines q_1, q_2 and q_3, q_4 are themselves normals—that is, intersect at right angles. Let the common normal of q_1 and q_2 be q_5, and that of q_3, q_4 be q_6. Apply the formula § 55 (3) to the sets

$$\begin{matrix} q_1 & q_2 & q_5 & q_6 \\ q_3 & q_4 & q_5 & q_6 \end{matrix}$$

Then

$$\begin{vmatrix} q_{13} & q_{14} & 0 & q_{16} \\ q_{23} & q_{24} & 0 & q_{26} \\ q_{53} & q_{54} & q_{55} & 0 \\ 0 & 0 & 0 & q_{66} \end{vmatrix} = 0$$

that is

$$q_{13}q_{24} = q_{23}q_{14}$$

or

$$\cos v_{13} \cos v_{24} = \cos v_{23} \cos v_{14}$$

So if the lines J_{13} and J_{24} are normal

$$\cos v_{12} \cos v_{34} = \cos v_{23} \cos v_{14}$$

If both happen, then

$$\cos v_{12} \cos v_{34} = \cos v_{13} \cos v_{24}$$

and hence the lines J_{23}, J_{14} are also normal. The four lines q_1, q_2, q_3, q_4 are then said to be *orthic*, the conditions being

$$\cos v_{23} \cos v_{14} = \cos v_{13} \cos v_{24} = \cos v_{12} \cos v_{34}$$

The four orthic lines and the six common normals form a

regular configuration of ten lines, each normal to three others. The ten are conveniently named by the symbols 0, 1, 2, 3, 4 taken two at a time (fig. 35 (*a*)). Then the lines $q_{\alpha\beta}$, $q_{\gamma\delta}$ are normal, and the lines $q_{\alpha\beta}$, $q_{\alpha\gamma}$, $q_{\alpha\delta}$, $q_{\alpha\varepsilon}$ are an orthic set. There are five orthic sets in the configuration.

The figure is that of a rectangular hexagon, the three common normals of opposite sides, and the common normal

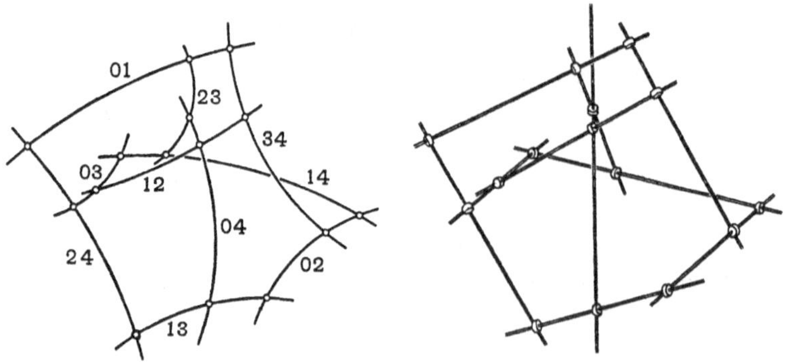

(a) (b)

Fig. 35

of these three. There are in it ten rectangular hexagons. Fig. 35 (*b*) is a sketch of the model for the parabolic case.*

The orthic line of three given lines A_i answers to the orthocentric point of the Euclidean triangle. The triangle *a*, *b*, *c* with its altitudes meeting at *d* and the normals to the plane at *a*, *b*, *c*, *d* is in fact a special case.

Exercise 5 – The configuration of ten quadratics may be taken as

$$\begin{array}{llll} 2x, & x^2-2\kappa x/(b+c)+a, & x^2-2\kappa' x/(b+c)+a, & x^2-a \\ & x^2-2\kappa x/(c+a)+b, & x^2-2\kappa' x/(c+a)+b, & x^2-b \\ & x^2-2\kappa x/(a+b)+c, & x^2-2\kappa' x/(a+b)+c, & x^2-c \end{array}$$

where

$$2\kappa\kappa'=(b+c)(c+a)(a+b)$$

The orthic four-line q_i may be defined by the fact that constants λ_i may be found such that

$$\lambda_1 q_1{}^2+\lambda_2 q_2{}^2+\lambda_3 q_3{}^2+\lambda_4 q_4{}^2=0$$

* In the restricted case of real numbers, the configuration becomes the Desargues configuration.

Taking any three quadratics there is the relation

$$\begin{vmatrix} q_{11} & q_{12} & q_{13} & q_1 \\ q_{21} & q_{22} & q_{23} & q_2 \\ q_{31} & q_{32} & q_{33} & q_3 \\ q_1 & q_2 & q_3 & 0 \end{vmatrix} \equiv 0$$

For in the relation D_4, $|q_{i1}\,q_{i2}\,q_{i3}\,q_{i4}| = 0$, let q_4 be a square, $(x-y)^2$. Then q_{44} is 0 and q_{14} is $a_1y^2 + 2\beta_1 y + \gamma_1$.

Thus D_4 becomes the above relation, with y in place of x.

Developed, the relation is, say

$$a_{11}q_1^2 + a_{22}q_2^2 + a_{33}q_3^2 + 2a_{23}q_2q_3 + 2a_{31}q_3q_1 + 2a_{12}q_1q_2 = 0$$

or

$$(a_{11}/a_{23}a_{31}a_{12} - 1/a_{23}^2)q_1^2 + \ . \ . \ . + (q_1/a_{23} + q_2/a_{31} + q_3/a_{12})^2 = 0$$

The three quadratics q_1, q_2, q_3 then determine a unique fourth such that

$$\lambda_1 q_1^2 + \lambda_2 q_2^2 + \lambda_3 q_3^2 + \lambda_4 q_4^2 \equiv 0$$

for any value of the variable x.

Now $\lambda_1 q_1^2 + \lambda_2 q_2^2 = 0$ is four points giving a pair of lines normal to the common normal of q_1 and q_2, j_{12}. But then also $\lambda_3 q_3^2 + \lambda_4 q_4^2 = 0$, the same four points giving a different pair of lines normal to j_{34}. By the theory of four points the lines j_{12} and j_{34} are then normal.

§ 58. **Geometry of the Three-line** – We have had the generalisation, to any three lines of a hyperbolic space, of the inscribed circle and the orthocentre of the Euclidean triangle. It is easy to carry the generalisation further. In terms of three given quadratics A_i any other is

(1) $$\lambda_1 A_1 + \lambda_2 A_2 + \lambda_3 A_3$$

It is a square if

(2) $$(\Omega\lambda)^2 \equiv \lambda_1^2 A_{11} + 2\lambda_2\lambda_3 A_{23} + \ . \ . \ . = 0$$

If we call (1) the line λ, and λ_i its homogeneous co-ordinates, then (2) is an equation of the sphere Ω. Two lines λ, μ have a vector v where

$$\cos v = (\Omega\lambda)(\Omega\mu)/\sqrt{(\Omega\lambda)^2}\sqrt{(\Omega\mu)^2}$$

where by $(\Omega\lambda)(\Omega\mu)$ we mean

$$\lambda_1\mu_1 A_{11} + (\lambda_2\mu_3 + \lambda_3\mu_2)A_{23} + \ . \ . \ .$$

the polarised form of (2).

But it would be better to consider the rectangular hexagon as the base. If a line X makes with the three A_i the vectors a_i and a line Y makes with three B_i the vectors β_i, then applying the formula D_4 to A_1, A_2, A_3, Y and B_1, B_2, B_3, X we have

$$\begin{vmatrix} c_1 & 0 & 0 & \cos a_1 \\ 0 & c_2 & 0 & \cos a_2 \\ 0 & 0 & c_3 & \cos a_3 \\ \cos \beta_1 & \cos \beta_2 & \cos \beta_3 & \cos v \end{vmatrix} = 0$$

so that the vector v of X, Y is given by

$$c_1c_2c_3 \cos v = \Sigma c_2c_3 \cos a_1 \cos \beta_1$$

Thus if X is Y, and $a_1 = a_2 = a_3 = r$, $\beta_1 = \beta_2 = \beta_3 = r'$, then

$$c_1c_2c_3 = rr'\Sigma c_2c_3$$

the formula which connects the radii of the cylinders inscribed to A_i and to B_i. The cylinder inscribed to B_i may be taken as the generalisation of the circumcircle of the Euclidean triangle $A_1A_2A_3$.

§ 59. **Rectangular Axes** – We may take as fundamental quadratics the three

$$q_1 = 2\iota x$$
$$q_2 = \iota(x^2 - 1)$$
$$q_3 = x^2 + 1$$

so written that

$$q_1{}^2 + q_2{}^2 + q_3{}^2 = 0$$
$$q_{11} = q_{22} = q_{33} = 1$$
$$q_{23} = q_{31} = q_{12} = 0$$

The lines intersect at right angles. They are rectangular axes.

Any quadratic is now

$$q_4 = \lambda_1 q_1 + \lambda_2 q_2 + \lambda_3 q_3$$

and

$$q_{14} = \lambda_1$$

so that

$$\cos v_{14} = q_{14}/\sqrt{q_{44}}\sqrt{q_{11}}$$

We take then λ_i as the cosines of the vectors made with the axes.

The fundamental relation, § 54 (2) becomes

$$\begin{vmatrix} 1 & 0 & 0 & \lambda_1 \\ 0 & 1 & 0 & \lambda_2 \\ 0 & 0 & 1 & \lambda_3 \\ \lambda_1 & \lambda_2 & \lambda_3 & 1 \end{vmatrix} = 0$$

or

$$\lambda_1^2 + \lambda_2^2 + \lambda_3^2 = 1$$

Applied to two lines λ and μ the relation § 54 (3) gives

$$\begin{vmatrix} 1 & 0 & 0 & \lambda_1 \\ 0 & 1 & 0 & \lambda_2 \\ 0 & 0 & 1 & \lambda_3 \\ \mu_1 & \mu_2 & \mu_3 & \cos(\lambda, \mu) \end{vmatrix} = 0$$

or

$$\cos(\lambda, \mu) = \lambda_1\mu_1 + \lambda_2\mu_2 + \lambda_3\mu_3$$

Exercise 6 – The Jacobian of the quadratics λ and μ is

$$| q_i \lambda_i \mu_i |$$

§ **60. The Euclidean Case** – To apply what has been said in this chapter to the lines of Euclidean space, we replace each vector $(\theta - \iota\lambda)$ by $(\theta - \iota\lambda/\mathrm{R})$, where R is the radius of Ω, and take the constant term and the coefficient of $1/\mathrm{R}$ of the equation considered as R tends to ∞. Then θ becomes the angle of the lines, and λ becomes their Euclidean distance. Each equation gives in this way two equations in reals. The first is that obtained by omitting the logarithmic distances λ, so that only angles enter. The second may be derived from this by the operator $\Sigma\lambda_i d\theta_i$.

Thus for the rectangular hexagon, where we write

$$a_i = \theta_i - \iota l_i$$
$$b_i = \phi_i - \iota m_i$$
$$c_i = \psi_i - \iota n_i$$

the equation

$$\cos c_1 \sin a_1 = \sqrt{(1 - \Sigma \cos^2 a_1 + 2 \cos a_1 \cos a_2 \cos a_3)}$$

gives first,

$$\cos \psi_1 \sin \theta_1 = \sqrt{(1 - \Sigma \cos^2 \theta_1 + 2 \cos \theta_1 \cos \theta_2 \cos \theta_3)}$$

(1) $$= \sqrt{\mathrm{A}}$$

and secondly,

$$l_1 \cos \psi_1 \cos \theta_1 - n_1 \sin \psi_1 \sin \theta_1$$
$$= [\Sigma l_1(\cos \theta_1 \sin \theta_1 - \sin \theta_1 \cos \theta_2 \cos \theta_3)]/\sqrt{A}$$
$$= [\sin \theta_1 \sin \theta_2 \sin \theta_3 \Sigma l_1 \cos \phi_1]/\sqrt{A}$$

so that

$$(2) \quad l_1 \cot \theta_1 - n_1 \tan \psi_1 = [\sin \theta_1 \sin \theta_2 \sin \theta_3 \Sigma l_1 \cos \phi_1]/A$$

These, with the equations obtained by interchanging θ, l with ϕ, m, are what the law of sines becomes for the rectangular hexagon in a Euclidean space.

Again, the equations for four orthic lines

$$\cos v_{23} \cos v_{14} = \cos v_{31} \cos v_{24} = \cos v_{12} \cos v_{34}$$

give for the Euclidean case, first,

$$\cos \theta_{23} \cos \theta_{14} = \cos \theta_{31} \cos \theta_{24} = \cos \theta_{12} \cos \theta_{34}$$

and, operating on the logarithms of these,

$$\lambda_{23} \tan \theta_{23} + \lambda_{14} \tan \theta_{14} = \lambda_{31} \tan \theta_{31} + \lambda_{24} \tan \theta_{24}$$
$$= \lambda_{12} \tan \theta_{12} + \lambda_{34} \tan \theta_{34}$$

In some cases we may equate the coefficients of $1/R^2$. Thus, in this case, where

$$\cos(\theta_{23} - \iota\lambda_{23}/R) \cos(\theta_{14} - \iota\lambda_{14}/R) = \ldots = \ldots$$

we have for Euclidean parallel lines

$$\cosh \lambda_{23}/R \cosh \lambda_{14}/R = \ldots = \ldots,$$

whence

$$\lambda_{23}^2 + \lambda_{14}^2 = \lambda_{31}^2 + \lambda_{24}^2 = \lambda_{12}^2 + \lambda_{34}^2$$

the elementary relations on orthic points.

Exercise 7 – In Euclidean space a cylinder touches three lines, making the same angle in each case. Find its radius.

§ 61. **Six Perpendicular Lines** – Anticyclic pairs are typified by 0, ∞ and at, a/t. For these pairs the vector $\theta - \iota\lambda$ is given by

$$\cos(\theta - \iota\lambda) = (t + 1/t)(1/t - t)$$

a pure imaginary. Hence $\theta = \pi/2$. The lines of the hyperbolic space are then perpendicular.

All lines q_i perpendicular to 0, ∞, or q_0 are thus given by

$\rho_i\tau_i t_i$, $\rho_i\tau_i/t_i$. If two of these, q_1 and q_2, are themselves perpendicular, then

$$\frac{\rho_1\tau_1(t_1+1/t_1)\rho_2\tau_2(t_2+1/t_2)-2\rho_1^2\tau_1^2-2\rho_2^2\tau_2^2}{\rho_1\tau_1(t_1-1/t_1)\rho_2\tau_2(t_2-1/t_2)}$$

is an imaginary $\iota\eta$, whence

$$\rho_1\tau_1(t_1+1/t_1)\rho_2\tau_2(t_2+1/t_2)=(\rho_1^2+\rho_2^2)(\tau_1^2+\tau_2^2)$$

Writing for shortness $\rho_i^2=\lambda_i$, $\tau_i^2=\mu_i$, if we take four lines q_i perpendicular to q_0 and to each other we have

$$\begin{aligned}(\lambda_2+\lambda_3)(\lambda_1+\lambda_4)(\mu_2+\mu_3)(\mu_1+\mu_4)&\\ &=(\lambda_3+\lambda_1)(\lambda_2+\lambda_4)(\mu_3+\mu_1)(\mu_2+\mu_4)\\ &=(\lambda_1+\lambda_2)(\lambda_3+\lambda_4)(\mu_1+\mu_2)(\mu_3+\mu_4)\\ &=\kappa, \text{ say.}\end{aligned}$$

Hence

$$\begin{aligned}(\lambda_2-\lambda_3)(\lambda_1-\lambda_4)&=\kappa/(\mu_3+\mu_1)(\mu_2+\mu_4)-\kappa/(\mu_1+\mu_2)(\mu_3+\mu_4)\\ &=-\kappa(\mu_2^2-\mu_3^2)(\mu_1^2-\mu_4^2)/\pi\end{aligned}$$

where π is the product of the six factors $(\mu_1+\mu_2)$, . . .

The λ_i are homographic with the μ_i^2, and similarly the μ_i are homographic with the λ_i^2. Hence we are concerned with five values of λ_i and of μ_i, and starting with three pairs of points q_i we obtain symmetrically five pairs of points—six pairs in all. Thus we have a set of six lines any two of which are perpendicular. Taking four perpendicular lines q_0, q_1, q_2, q_3, then there are two lines q_4, q_5 perpendicular to the four; and these two lines are themselves perpendicular.

This theorem is the inversive equivalent of the existence of the double-six of lines in a projective space, and the algebra is adapted from the paper of G. T. Bennett on the Double-six.* Strictly the theorem should be here proved also for an elliptic space, but we can rely on the argument of § 49.

§ 62. **The Vector of Two Directed Cylinders** – We have associated with each homography H on the sphere Ω a directed cylinder. We may call

$$\mathrm{H}\equiv\alpha xy+\beta x+\gamma y+\delta=0$$

the equation of the cylinder. Its axis is

$$\alpha x^2+(\beta+\gamma)x+\delta=0$$

---

* *Proc. London Math. Soc.*, series 2, vol. 9, p. 336.

and its vector radius is given by

$$\cos^2 r/(\beta-\gamma)^2 = -\sin^2 r/4(\alpha\delta-\beta\gamma) = 1/[(\beta+\gamma)^2 - 4(\alpha\delta-\beta\gamma)]$$

Two cylinders H_1 and H_2 have the invariant

$$2H_{12} = \alpha_1\delta_2 + \alpha_2\delta_1 - \beta_1\gamma_2 - \beta_2\gamma_1$$

When it vanishes they are said to be apolar, or to have positive contact.

The cylinder H_1 has the invariant

$$H_{11} = \alpha_1\delta_1 - \beta_1\gamma_1$$

When this invariant vanishes the cylinder is self-apolar, or singular. It is then Ω itself with two points marked on it.

We have then for two cylinders an absolute invariant

$$\cos v_{12} = H_{12}/\sqrt{H_{11}}\sqrt{H_{22}}$$

and we call v_{12} the vector of the two cylinders.

To see the meaning of v_{12} we consider again the rectangular hexagon (fig. 33). Let A_1 be the axis, and b_2 the radius, of the first cylinder; and let A_2 be the axis, and b_1 the radius, of the second. It is clear that a_3 is a common tangent of the two cylinders which agrees with each in direction. We call this a positive common tangent. Are not these two vectors, a_3 geometrically given and v_{12} analytically given, the same?

We have

$$\cos b_2/(\beta_1-\gamma_1)^2 = \iota \sin b_2/2\sqrt{H_{11}} = 1/\sqrt{(\beta_1+\gamma_1)^2 - 4H_{11}}$$

$$\cos b_1/(\beta_2-\gamma_2)^2 = \iota \sin b_1/2\sqrt{H_{22}} = 1/\sqrt{(\beta_2+\gamma_2)^2 - 4H_{22}}$$

$$\cos v_{12} = \frac{1}{2}\,\frac{\alpha_1\delta_2 + \alpha_2\delta_1 - \beta_1\gamma_2 - \beta_2\gamma_1}{\sqrt{\alpha_1\delta_1 - \beta_1\gamma_1}\sqrt{\alpha_2\delta_2 - \beta_2\gamma_2}}$$

so that

$$\cos v_{12} + \cot b_1 \cot b_2 = \frac{2(\alpha_1\delta_2 + \alpha_2\delta_1) - (\beta_1+\gamma_1)(\beta_2+\gamma_2)}{4\sqrt{H_{11}}\sqrt{H_{22}}}$$

But the vector b_3 of the axes is given by

$$\cos b_3 = \frac{2(\alpha_1\delta_2 + \alpha_2\delta_1) - (\beta_1+\gamma_1)(\beta_2+\gamma_2)}{\sqrt{(\beta_1+\gamma_1)^2 - 4H_{11}}\sqrt{(\beta_2+\gamma_2)^2 - 4H_{22}}}$$

Hence

$$\cos v_{12} + \cot b_1 \cot b_2 = \cos b_3/\sin b_1 \sin b_2$$

or

$$\cos v_{12} \sin b_1 \sin b_2 = \cos b_3 - \cos b_1 \cos b_2$$

Thus it is true that

$$a_3 = v_{12} \pmod{2\pi}$$

Thus the vector v_{12} of two directed cylinders is shown by a positive common tangent. There are two such tangents, but they have the same vector. It is to be noticed that $H_1H_2^{-1}$, where, as here, we are speaking of general homographies, must have fixed points, so that the common tangents exist.

§ 63. **The Relation D_5** – We have between any five homographies H_i a linear relation, obtained by eliminating xy, x, y, 1. Hence, as in § 54, there is the relation

(1) $$D_5 \equiv | \cos v_{ij} | = 0$$

and more generally between two sets of five homographies H_i and $H_{i'}$ there is the relation

(2) $$D_5 \equiv | \cos v_{ii'} | = 0$$

If, then, four cylinders touch a fifth positively we have, since $\cos v_{i5} = 1$,

$$\begin{vmatrix} 1 & \cos v_{12} & \cos v_{13} & \cos v_{14} & 1 \\ \cos v_{21} & 1 & \cos v_{23} & \cos v_{24} & 1 \\ \cos v_{31} & \cos v_{32} & 1 & \cos v_{34} & 1 \\ \cos v_{41} & \cos v_{42} & \cos v_{43} & 1 & 1 \\ 1 & 1 & 1 & 1 & 1 \end{vmatrix} = 0$$

or, writing s_{ij} for $\sin v_{ij}/2$,

$$\begin{vmatrix} 0 & s_{12}^2 & s_{13}^2 & s_{14}^2 \\ s_{21}^2 & 0 & s_{23}^2 & s_{24}^2 \\ s_{31}^2 & s_{32}^2 & 0 & s_{34}^2 \\ s_{41}^2 & s_{42}^2 & s_{43}^2 & 0 \end{vmatrix} = 0$$

or

(3) $$s_{23}s_{14} \pm s_{31}s_{24} \pm s_{12}s_{34} = 0$$

If, then, four cylinders touch positively a fifth, we have, say,

(4) $$s_{23}s_{14} + s_{31}s_{24} + s_{12}s_{34} = 0$$

If we reverse the directions of two of the cylinders, say H_2 and H_3, then we must replace the direct vector of H_1 and H_2 by the transverse vector, and s_{12} by, say, s_{12}'. Let then

$$s_{23}s_{14} = s_{31}'s_{24}' - s_{12}'s_{34}'$$

and when we reverse the directions of H_3 and H_1 let

$$s_{31}s_{24} = s_{12}'s_{34}' - s_{23}'s_{14}'$$

Then we have by addition

$$s_{12}s_{34} = s_{23}'s_{14}' - s_{31}'s_{24}'$$

If, then, four cylinders touch a fifth in the sense $++++$, and touch a sixth in the sense $+--+$, and touch a seventh in the sense $-+-+$, they can touch an eighth in the sense $--++$.

That is, Hart's extension of Feuerbach's theorem is true for cylinders. For the accurate handling of Hart's theorem we refer to Study.*

§ 64. **The Caustic of a Correspondence** – We suppose now a correspondence on Ω between the points x and y, and that this is differentiable, so that for $x+dx$ we have in general $y+dy$.

We have a double infinity or congruence of arcs x, y. And the problem is to determine their envelope. This is, in a Euclidean space (when Ω is a point), a part of Hamilton's theory of systems of rays.

It is convenient to regard the sphere Ω as a plane. An arc x, y is then a semicircle, say above the plane. The arc x, y intersects a consecutive arc $x+dx$, $y+dy$ when the pairs of points are on a circle and are interlaced. The cross-ratio $-dxdy/(x-y)^2$ is then negative. We have then, when $d\theta$ is real,

$$dxdy/(x-y)^2 = (d\theta)^2 \tag{1}$$

Thus the condition of intersection is that this differential invariant under homographies is to be invariant also under antigraphies (that is, not altered by writing $\bar{x}$ for x) and is further to be positive. Geometrically stated, the elements dx and dy make opposite angles with $x-y$.

To find the point of intersection, we first find the point z where the join of x and y meets the join of $x+dx$ and $y+dy$. Taking the circle on which they lie as the base-circle, and replacing x and y by turns t and τ we are to have

$$z + \bar{z}t\tau = t + \tau$$

* "Das Apollonische Problem," *Math. Annalen*, v. 49 (1897).

and

$$\bar{z}(\tau dt + td\tau) = dt + d\tau$$

Let $dt/d\tau = \mu e^{\iota a}$, where μ is positive (the magnification).

Then $\tau^2 dt/t^2 d\tau = \mu e^{\iota a}$

$$dt/d\tau = \pm \mu t/\tau$$

For interlaced pairs, dt/t and $d\tau/\tau$ have the same sign, so that, as we have taken μ positive,

$$dt/d\tau = \mu t/\tau$$

We have then

$$\bar{z}(1+\mu) = 1/t + \mu/\tau$$
$$z(1+\mu) = t + \mu\tau$$

and therefore in general

(2) $$z = (x + \mu y)/(1+\mu)$$

where

$$\mu = | \, dx/dy \, |$$

The point z is thus the internal centre of similitude for corresponding small circles around x and y.

On the normal to Ω at this point z, we take a distance ζ given by

(3) $$\zeta^2 = (z - x)(\bar{y} - \bar{z})$$

to obtain the intersection of the consecutive arcs. The envelope of the arcs is then given by (1), (2), and (3). For convenience let it be called the *caustic* of the correspondence.

This applied to the homography $\kappa^2 x = y$ will give a right cone, as we saw directly in § 53.

Consider the antigraphy

$$a x\bar{y} + \beta x + \gamma \bar{y} + \delta = 0$$

When there are fixed points, say 0 and ∞, this is

$$x = \mu \bar{y}$$

The equations (1), (2), (3) are unaltered. And from (1),

$$\frac{\mu dy d\bar{y}}{(x-y)^2} = (d\theta)^2$$

Hence $x - y$ is a real.

Hence x, y, and z are reals. Thus in this case the only arcs which can intersect consecutive arcs are those erected on the axis of reals. We have

$$\zeta^2 = (1-\mu)^2 z^2/4\mu$$

where z is a real; and the arcs touch two half-lines in the vertical plane on the axis of reals, the lines being images as to Ω. For any antigraphy with fixed points, the envelope of arcs is accordingly two arcs which meet Ω at the fixed points.

When the antigraphy has interchanging points 0, ∞, it may be taken

$$x\bar{y} = t^2$$

where t is a given turn.

From (1) we have

$$(d\theta)^2 = -\frac{x}{y}\frac{dyd\bar{y}}{(x-y)^2}$$

so that

$$\frac{x}{\bar{y}(x-y)^2} = \frac{\bar{x}}{y(\bar{x}-\bar{y})^2}$$

that is, $xy/(x-y)^2$ is a real.

Hence x and therefore y must be taken on the base-circle. The envelope of the arcs is then a circle parallel to Ω, lying on the base-sphere.

For an antigraphy with interchanging points, the envelope of arcs is accordingly a circle, such that the spheres on it, which touch Ω, touch it at the interchanging points.

In general for an algebraic correspondence $f(x, y) = 0$, that is for a Riemann surface, we shall have a caustic *surface*. The equation (1) giving at the place (x, y), or its correspondent (y, x), two directions at right angles, gives a natural dividing of the surface into orthogonal curves.

But for a correspondence $f(x, \bar{y}) = 0$, we shall have a caustic *curve*. In particular this is the case when the equation is self-conjugate—that is, when it is the image-system of an algebraic curve.

For antigraphies in the Euclidean case, the circles which cut Ω, other than orthogonally, disappear. We are left with a circle of the Euclidean space.

CHAPTER X

FLOW

§ 65. **Analytic Flow** – In moving a card over a table (§ 1) we may consider a particle at $\xi + \iota\eta$. If we first fix ξ we have a line in the card, for varying η. If we then vary ξ continuously, the line moves. This continuous motion is a rigid motion. If we first fix η, varying ξ, and then vary η we have a second rigid motion. It is orthogonal (or conjugate) to the first.

If we call a particle ρt, then if we fix t we have a ray in the card. If then we vary t continuously we have a continuous rotation. This again is a rigid motion. If we first fix ρ, we have a circle in the card; and when we vary ρ continuously we have a continuous motion—a fluid motion, or flow. It is orthogonal to the rotation. Let us call it radial flow. The points 0 and ∞ are here exceptional. They are excluded points, or punctures in the plane. They are called point-charges, or a sink and an equal source. We associate the number z, written as ρt, with radial flow.

An analytic function of z will then give another flow. The rays and circles give the paths and levels. This kind of flow, set up by analytic functions, is called the flow of a perfect fluid. The best illustration is electricity. Here, of course, what is called geometrical drawing—with rigid instruments—is secondary. What is primary is electrical drawing. The process (originated by Kirchhoff) is described in its simplest form in Ames and Bliss, *Manual of Experiments in Physics.* Suppose, for instance, a rectangular tray containing a conducting fluid, and in it a circular disc. A flow is set up between the disc and the rectangle, and by means of a telephone the levels can be found. The transition from the circle to the rectangle is put in evidence.

In the radial flow it is clearly convenient to consider the

region within a circle, say the base-circle. The flow is then from the centre to the circle. If we choose our analytic function so that there are no singular points in this circular region, then it maps into a new region R of simple character and we shall have the flow in the region R from a source to the rim. The function sends the interior of the base-circle into R, but conversely it has been shown by long and keen concentration, principally German,* that the region R defines the function, subject to the homographies which send the base-circle into itself. This amounts to a definition of the shape of a region, when there is one rim.

Instead of the circular region we may use the half-plane as the base-region, say B. We have for the circle or the half-plane a hyperbolic geometry and an element of distance, or metric. For the half-plane the metric is ds/η when ds is the Euclidean element and η the ordinate. We have the geodesics or paths of least distance. In the region R we have then a hyperbolic geometry. It is convenient to carry over the element of distance, unchanged. It is then obvious that the geodesics of B become geodesics of R. And as in B the paths of flow from a point to the rim are along geodesics, so in R.

Exercise 1 – The function $x = \log z$ sends the upper half-plane of z into the strip $0 \leqslant \eta \leqslant \pi$, where $x = \xi + \iota\eta$. It sends the geodesic $z\bar{z} - \mu(z+\bar{z}) + \kappa = 0$ into the geodesic

$$e^{x+\bar{x}} - \mu(e^{x} + e^{\bar{x}}) + \kappa = 0$$

that is

$$e^{\xi} - 2\mu \cos \eta + \kappa e^{-\xi} = 0$$

Draw the paths of flow (that is the geodesics) from a point within the strip to the rim.

§ 66. **Standard Case of Radial Flow** – We regard then the process of mapping as giving the *flow* of a perfect fluid. Let us set up first the standard case of flow, and consider a few simple cases which are derived from the standard case.

In a horizontal plane covered with a layer of fluid a *sink* or small circular hole is made. The fluid flows radially towards this sink. To balance the loss we suppose that an equal amount is continually supplied by a *source* at infinity.

* See § 80 for some references.

The velocity of the fluid at a distance ρ from the sink is determined by the condition that the amount which crosses the circle of radius ρ is independent of ρ. That is, denoting the velocity by $\dot{\rho}$,

$$2\pi\rho\dot{\rho} = \text{constant}$$

or

$$\dot{\rho} = \kappa/\rho$$

The proper co-ordinates here are ρ and t, and the expression of the flow is

$$z = \rho t$$

Thus the polar grid is associated with the flow of a fluid due to a sink at 0 and an equal source at ∞. The source at ∞ need cause no difficulty; the point we wish to make is that the operation of mapping amounts merely to changing the conditions of the flow. On the globe, from which is derived the polar grid by an inversion, the source is conceived at (say) the north pole and the equal sink at the south pole, the fluid flowing over (say) the outer surface from north to south.

Exercise 2 – The velocity along the axis of the globe is constant.

We take this case of radial flow to be the standard case of flow. In any isogonal mapping,

$$z = f(x)$$

the standard flow maps into a flow in the x-plane, with the assumption that the sink ($z=0$) gives sinks in the x-plane, and that the source ($z=\infty$) gives sources. Then the lines $t=$ constant will map into the curves of flow corresponding to the sources and sinks in the x-plane, and the concentric circles $\rho=$ constant will map into the curves of level for the flow in the x-plane.

§ 67. **Two Equal Sinks** – Suppose the map-equation

$$\begin{aligned} z &= \text{quadratic in } x \\ &= (x - x_1)(x - x_2) \\ &= \rho_1\rho_2 t_1 t_2 \end{aligned}$$

We have then in the x-plane two sinks, supposedly congruent.

To rays on the base-point in the standard case (z-plane) correspond the curves

$$t_1 t_2 = \text{constant}.$$

Therefore a particle x moves so that the product of its directions from the two sinks is constant; that is,

$$\theta_1 + \theta_2 = \text{constant},$$

where θ_i is the angle made by $(x - x_i)$ with any fixed line. The curves of flow are rectangular hyperbolas (fig. 36).

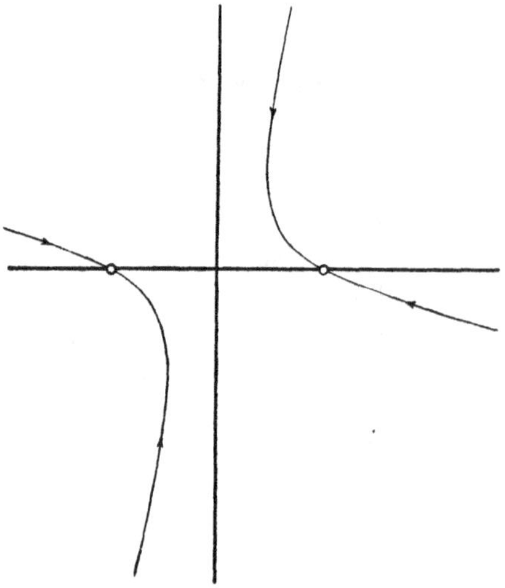

FIG. 36

The curves of level, given by

$$\rho_1 \rho_2 = \text{constant}$$

are, when the constant is small, little ovals surrounding each sink. It is convenient to speak of sinks (and sources) as points; but if in deference to physical interpretation one wishes them to have finite size, we take each sink to be the interior of one of these ovals. To balance the loss we suppose again a source at ∞ of capacity equal to the sum of the two sinks.

Here we have flow from ∞ to either x_1 or x_2. The separating curve is always especially to be noted; here it is the axis of x_1, x_2 together with the line on x_1, x_2. The hyperbola which separates the flow is formed by these two perpendicular lines.

In general when

$$z = f(x)$$

there will be values of z for which two values of x become equal. These are *branch-points* in the z-plane, and the corresponding positions of x—the positions where two x's come together—are in Klein's nomenclature the *crossing-*

points. Thus to $z=b_i$ corresponds $x=c_i$. To a first approximation we have

$$z-b_i=A(x-c_i)^2$$

The isogonality breaks down; an angle at b_i is twice the corresponding angle at c_i. Thus to the ray on b_i in the standard flow corresponds a curve cutting itself at right angles at c_i. The crossing-points are characterised by the vanishing of $D_x z$, or may also be regarded as the Jacobian points of all sinks and all sources. The curves of flow which pass through the crossing-points are the separating curves, and in complicated cases give essential help in visualising the flow. In any figure of flow they should be especially noted or sought.

§ 68. **Three Equal Sinks** – Let

$$z=\text{cubic in } x$$

The crossing-points are given by $D_x z=0$, a quadratic. For simplicity take these points as ± 1, so that (say)

$$D_x z=3(x^2-1)$$

Then the map-equation is of the form

$$z=x^3-3x+c$$

The curves $t=$ constant or $z/\bar{z}=$ constant become the cubics

$$\frac{x^3-3x+c}{\bar{x}^3-3\bar{x}+\bar{c}}=t^2$$

Or directly from the map-equation, written

$$\begin{aligned} z &= (x-x_1)(x-x_2)(x-x_3) \\ &= \rho_1 t_1 \cdot \rho_2 t_2 \cdot \rho_3 t_3 \end{aligned}$$

the curves of flow are such that

$$t_1 t_2 t_3=\text{constant}$$

or

$$\theta_1+\theta_2+\theta_3=\text{constant} \quad (\text{mod } 2\pi)$$

The separating curves are as shown in fig. 37, and neighbouring curves of flow are sketched in by their guidance.

By the same method cases of more numerous sinks may be

treated. For any polynomial $f(x)$ the curves of flow are given by

$$\Sigma\theta_i = \text{constant} \quad (\text{mod } 2\pi)$$

where θ_i is the angle of the vector from a sink to x; and the curves of level by

$$\rho_1\rho_2 \; . \; . \; . \; \rho_n = \text{constant}$$

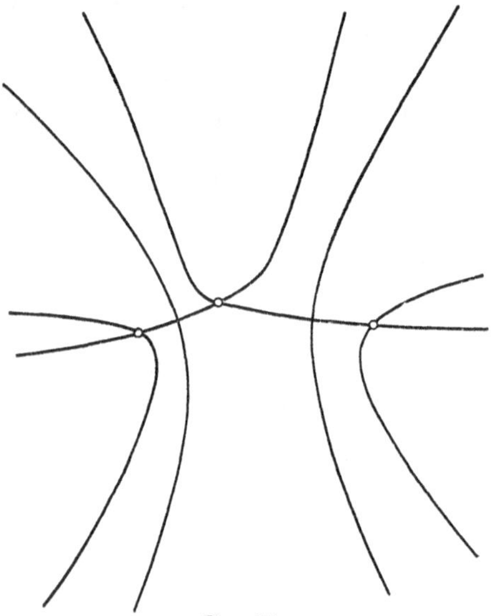

FIG. 37

where ρ_i is the distance from a sink to x.

Unequal sinks are dealt with by allowing sinks to coincide when their strengths are commensurable. When they are not commensurable a limiting process is necessary.

§ 69. The Rational Fraction – For mapping with the rational fraction, consider

$$z = (x - x_1)/(x - x_2)$$

In the x-plane there is no longer a source at ∞; the source is at x_2 and is supposed equal to the sink at x_1. The flow is along arcs of circles, forming a hyperbolic pencil, from x_2 to x_1. The curves of level form the elliptic pencil, orthogonal to the curves of flow.

So generally when $z = f(x)$, a rational fraction, we have sources at all poles, sinks at all zeros, and unless they occur in equal numbers, a source or sink at ∞ to make the total the same. In general the curves of flow may be written

$$\Sigma\theta_i - \Sigma\phi_i = 0$$

where θ_i is the angle of the ray from a sink to x and ϕ_i the angle of the ray from a source to x.

The canonical case of two sinks and two sources is illustrated in fig. 38.

The simplicity so far is due to the fact that for one x we have only one z, so that there are on the plane of x no branch points.

§ 70. **The Doublet** – The case of coincident source and sink is important. For single source and single sink, given by

$$z=(x-x_1)/(x-x_2)$$

it is geometrically obvious that when x_2 and x_1 are allowed to coincide in a definite direction, the curves of flow become tangent circles.

Suppose both source and sink as of infinite strength — that is, as indefinitely repeated. To take the matter in simplest form we return to the standard radial flow. In this we write

$$z=(x-a)^n$$

so that the sink $z=0$ shall map not into a fixed point in the x-plane, but into a moveable point a. To arrange a relation between a and n so as to get a limit when $n \to \infty$, we should write

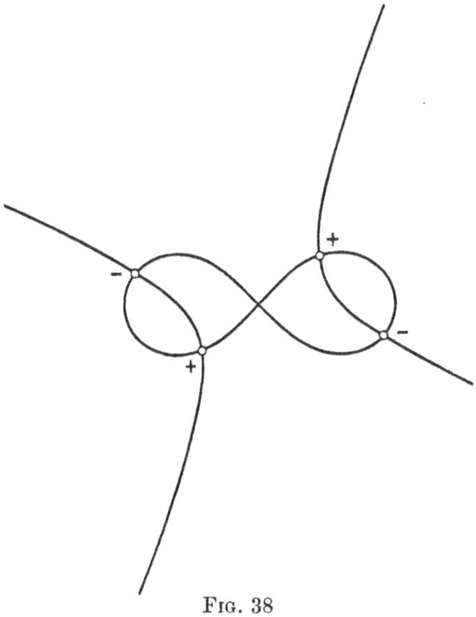

FIG. 38

$$z=(1+x/n)^n$$

Thus

$$z=\exp x$$

is the proper mapping for source and sink, both at infinity.

When source and sink unite in this way we have a doublet. Here the doublet is at ∞ in a given direction. The flow is that of a river between parallel lines. We take this parallel flow as the second standard case, mapped from the radial flow by the logarithm. As in § 12, in the x-plane the appropriate co-ordinates are the rectangular co-ordinates ξ, η,

$$\xi = \log \rho$$
$$\eta = \theta$$

But now we see that the point z is sent into a vertical row of points, since for θ we may write $\theta \pm 2n\pi$.

A doublet at a finite point a is analytically represented by

$$A/(x-a)$$

where A is constant. This is obtained directly from the equation

$$z = (x-b)/(x-a)$$

by letting b coincide with a in a given direction. Then, if

$$b - a = da$$
$$z = 1 - da/(x-a)$$

We replace this by

$$z = \left[1 - da/(x-a)\right]^n$$

and let

$$nda = -A$$

Then

$$z = \left[1 + \frac{1}{n} \cdot \frac{A}{(x-a)}\right]^n$$

and when $n \to \infty$,

$$z = \exp\,[A/(x-a)]$$

or

$$\log z = A/(x-a)$$

If $A = rt$, t is the direction of the doublet (or direction of flow at a) and the positive number r is the strength. Thus a doublet may be denoted as (a, rt).

The same process applies to any number of coincidences of sinks and sources. For n doublets we have the equation

$$\sum^{n} \frac{r_i t_i}{x - a_i} = \log z = \log \rho + \iota\theta$$

Hence the curves of flow, the maps of $\theta = \text{constant}$, will be

$$\sum \frac{r_i t_i}{x - a_i} - \sum \frac{r_i / t_i}{\bar{x} - \bar{a}_i} = 2\iota\theta$$

It is noticed that as sources and sinks call for the resolution of $f(x)$ into factors, so doublets call for $f(x)$ in partial fractions. We may of course write

$$y = \log z$$

and refer the general problem of the flow due to doublets (without additional sources and sinks) to parallel flow in the y-plane, this being the second case of standard flow.

§ 71. **Two Opposed Doublets** – We consider the simple example of two doublets

$$x + 1/x = 2 \log z$$

or

$$x + 1/x = 2y$$

where y is in parallel flow from right to left. The crossing-points are given by

$$1 - 1/x^2 = 0$$

and are $x = \pm 1$. As y describes the real axis from $+\infty$ to $+1$, x is real and decreases from $+\infty$ to $+1$. When y

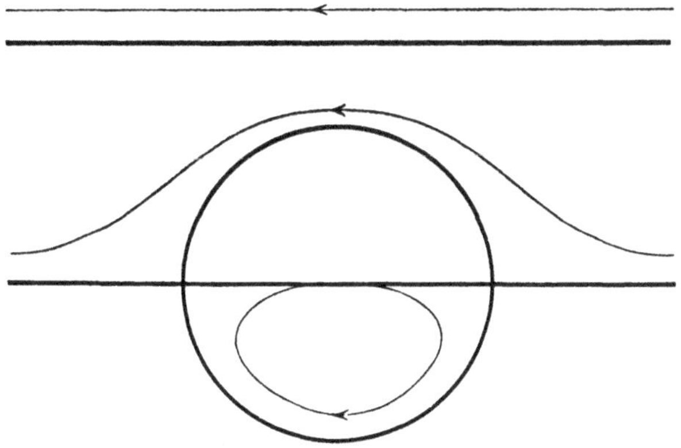

FIG. 39

describes the segment of the real axis from 1 to -1, x traces the base-circle. The separating curve is here the base-line and the base-circle. The curves of flow are the cubics

$$x + 1/x - \bar{x} - 1/\bar{x} = \iota\kappa$$

(fig. 39).

This is the problem of a circular island in parallel flow. It will be noticed that the doublets are opposed, in fact inverse as to the base-circle in the x-plane. Inversion, sending source to source and sink to sink, will send a doublet into a doublet, though the strength is in general altered. Normally, then, a case of flow due to doublets becomes under inversion another case. Here the case goes into itself. What is meant by opposed doublets is this interchangeability under inversion; they are elements of a circle and have opposite directions along that circle.

Exercise 3 – Let two doublets of the same strength 1, be $(1, \iota)$ and $(-1, -\iota)$. The curves of flow are

$$\kappa x^2\bar{x}^2 - (\kappa+1)(x^2+\bar{x}^2) + \kappa + 2 = 0$$

There are the special curves $\kappa=0$ (a rectangular hyperbola), $\kappa=-2$ (a lemniscate), $\kappa=-1$ (the base-circle together with $x\bar{x}+1=0$). This indicates the transformation $x\bar{y}=-1$ which, as well as $x\bar{x}=1$, will send the pencil of curves into itself.

We have said that the doublet corresponds to the simple partial fraction. If we have terms like

$$A/(x-a)^2$$

or, near ∞, a term

$$Ax^2$$

we regard this as the coincidence of doublets. The appropriate flow in the latter case is deduced from

$$x^2 = y$$

where the flow of y is parallel and again (say) from right to left. And similarly for Ax^n near ∞, or $A/(x-a)^n$ near a.

§ 72. **Flow with Doublets, Sinks, and Sources** – As a simple case, let there be a sink at 0 and a doublet at ∞. The source corresponding to the sink is supposed combined with the doublet. The case is that of parallel flow when a hole is made.

The appropriate equation, for flow from left to right, is

$$x \,.\, e^x = z$$

or

$$\log x - x = \log r + \iota\theta$$
$$\log \bar{x} - \bar{x} = \log r - \iota\theta$$

By subtraction

$$\tan^{-1}\eta/\xi - \eta = \theta$$

where $x = \xi + \iota\eta$, so that the curves of flow corresponding to $\theta = \kappa$ are

$$\eta = \xi \,.\, \tan(\eta + \kappa)$$

There is a finite crossing-point, for $D_x z = 0$ when $x = 1$. For this $\kappa = 0$, so that the separating curve is

$$\eta = \xi \tan \eta$$

of which the axis $\eta = 0$ is a part. Fig. 40 shows the flow.

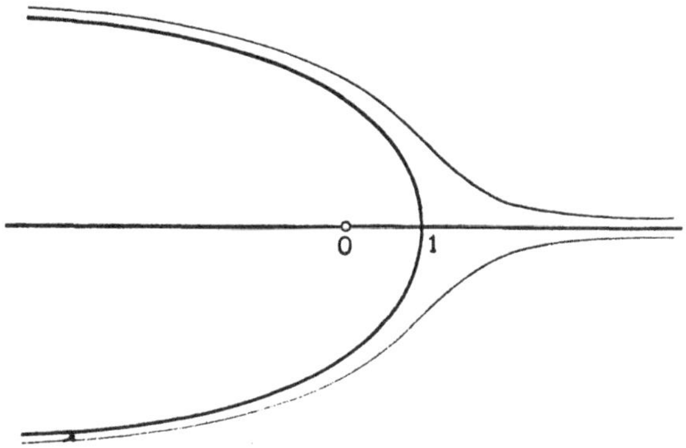

FIG. 40

Exercise 4 – By considering the orthogonal system, determine the shape of the hole or waterfall when not small.

In general the equation for coexisting doublets, sinks, and sources has the form

$$f(x) \exp g(x) = z$$

where the zeros of $f(x)$ are the sinks, the infinities of $f(x)$ the sources, and the infinities of $g(x)$ the doublets. Higher doublets may of course enter.

§ 73. **Vortices** – So far the levels, orthogonal to the curves of flow or paths, have only been mentioned incidentally as suitable forms for sinks and sources, or in other terminology for electrodes. But paths and levels may be interchanged. In the Ptolemy flow in the two-plane, the interchange leads us to consider motion (rotation) along the concentric circles. Here the base-point is a *vortex*. There is another vortex, here at ∞, said to be of opposite strength, or of same strength and opposite sign. A source or sink is a level which, usually speaking, reduces to a point; a vortex is a path which reduces to a point. Selected paths are often considered as boundaries, or as non-conductors; thus a vortex is a boundary which, usually speaking, reduces to a point. But without confusion the words sink and source are also extended to mean any level, and vortex to mean any boundary.

From the simplest case others are derived by writing

$$f(x)=z$$

The argument need not be repeated; for source is substituted vortex, for sink is substituted negative vortex. Factors $(x-a)$ and $(x-a)^{-1}$ are replaced by $(x-a)^{\iota}$ and $(x-a)^{-\iota}$, for the motion defined by

$$x^{\iota}=rt$$

for given t is

$$\iota(\log r_1+\iota\theta_1)=\log r+\iota\theta$$

or

$$\log r_1=\text{constant}$$

and this is rotation.

In particular two opposite vortices at a and b cause motion in circles, of which a and b are inverse points. Any two circles of the system may be regarded as vortices. This is the flow of a fluid if two circular (non-intersecting) islands are given, so long as there is no other agent present. When the points a, b coincide the circles touch and we have a doublet as in § 70.

It must be understood that for us vortices are fixed; the question of vortex filaments inherent in the fluid and capable of motion is not here considered.

The combined effect of sinks at points a_i and vortices at points b_i is given by

$$\Pi(x-a_i)^{\lambda_i}\Pi(x-b_i)^{\iota\mu_i}=rt$$

where λ and μ are the strengths. If ∞ be neither a vortex nor a source, then

$$\Sigma\lambda=0 \quad \text{and} \quad \Sigma\mu=0$$

If a vortex coincides with a sink, say at c, we have a factor

$$(x-c)^{\lambda+\iota\mu}$$

The equation

$$(x-c)^{\lambda+\iota\mu}=z$$

implies, in addition to the coincidence of vortex and sink at c, that the opposite vortex and source coincide at ∞. The paths are equiangular spirals.

As an example, let there be a sink at 0, vortex at 1, and the opposite vortex and source at ∞. We wish to draw the curve which separates the fluid which reaches 0 from that which whirls eternally round 1.

The curve is of the type

$$r_1 = e^{\kappa(\phi - a)}$$

where r_1 is the distance from the vortex and ϕ the angle from the sink. In the case

$$x(x-1)^{-\iota} = z$$

the crossing-points are given by

$$1/x - \iota/(x-1) = 0$$

or

$$x = 1/(1-\iota) = (1+\iota)/2$$

The curve is

$$r_1 \sqrt{2} = e^{\phi - \pi/4}$$

Thus having drawn

$$r = e^{\phi}$$

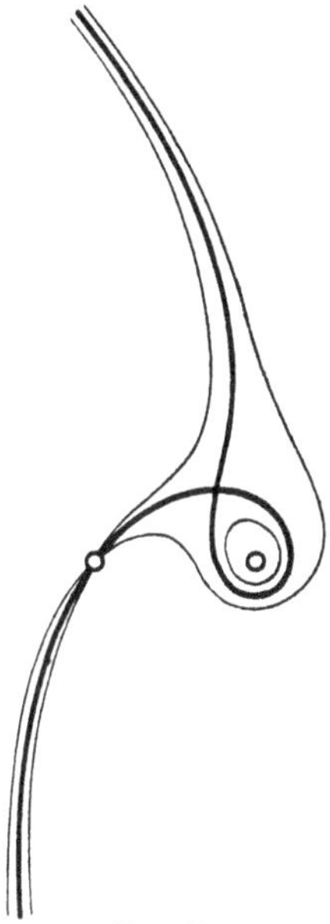

FIG. 41

a logarithmic spiral, to mark the points distant

$$r_1 = r/e^{\pi/4} . \sqrt{2}$$

from 1, we draw rays from 0 to the spiral and mark off on these rays points which are distant r_1 from the point 1. In fig. 41 the spiral has been erased.

CHAPTER XI

DIFFERENTIAL GEOMETRY

§ 74. The Translational Derivative – In the main we have so far considered algebraic processes. We here consider (though only a sketch is admissible) processes of differentiation and integration, and the so-called differential and integral invariants.

Thus for translations

$$y = x + b$$
$$dy = dx$$
$$d^2y/dx^2 = 0$$

For a plane curve,

$$dx = ds e^{\iota\phi}$$

where s is the length of arc from a point of the curve, ϕ the angle at the point. Since

$$d\bar{x} = ds e^{-\iota\phi}$$
$$dx d\bar{x} = ds^2$$

and

$$dx/d\bar{x} = e^{2\iota\phi}$$

is the clinant at the point. The curvature

$$c = 1/r \quad \text{is} \quad d\phi/ds$$

r being the radius of curvature.

Here the length s is an integral invariant, and the curvature a differential invariant, under translations.

The elementary operations of the calculus are built on the group of translations, as applied to a function. In the neighbourhood of a point x_0 a function y of x is defined by

$$y - y_0 = a_1(x - x_0) + a_2(x - x_0)^2/2\,! + \ . \ . \ .$$

where the series is convergent within some circle with centre x_0. Here $x - x_0$ is the change in x, $y - y_0$ is the change in y; if we denote these by ξ and η, the Taylor series is

(1) $$\eta = a_1\xi + a_2\xi^2/2\,! + \ . \ . \ . \ + a_n\xi^n/n\,! + \ . \ . \ .$$

This is also written

$$\eta = dy + d^2y/2\,! + \;.\;.\;.\; + d^ny/n\,! + \;.\;.\;.$$

and d^ny or $a_n\xi^n$ is called the nth differential of y (with respect to the independent variable x). For the independent variable there is only the first differential $\xi = dx$; all others are 0. When x_0 is 0, then dx and x are the same, and x is used, not dx. The coefficients a_n may be written as d^ny/dx^n; they are the differential coefficients or derivatives of y for the value x_0 of x. Notations in use for these coefficients are $D_x{}^ny$, or $f^n(x)$; in simple cases superposed dots, $\dot{y}$, $\ddot{y}$, etc.

When we wish to apply this conception of functions of a complex variable to curves, we take the equation of the curve as a map-equation by restricting x either to the axis of reals or to the base-circle; or else we consider the self-conjugate equation

$$f(x, \bar{x}) = 0$$

The curves arising thus from analytic functions are *analytic* curves. Usually in the second case the equation is taken to have only a finite number of terms. The curve is then *algebraic*.

The most useful method in general is to regard y as a function of a real variable or *parameter* ρ, which, in the first instance, is subject to translations only. We replace here y by x, so that the curve is

$$x - x_0 = f\,(\rho - \rho_0)$$

It is convenient to think of $\rho - \rho_0$ as the local time; then $\dot{x}$ is the velocity, $\ddot{x}$ the acceleration, of the moving point x.

Considering the neighbourhood of $\rho_0 = 0$, we have (replacing $x - x_0$ by x)

$$(2) \qquad x = f(\rho) = a_1\rho + a_2\rho^2/2\,! + a_3\rho^3/3\,! + \;.\;.\;.$$

If we have a second curve which intersects the first at the point $x = 0$, then for it, with the same parameter ρ,

$$x' = \beta_1\rho + \beta_2\rho^2/2\,! + \beta_3\rho^3/3\,! + \;.\;.\;.$$

If $a_1 = \beta_1$, the curves *touch* or have linear contact. If also $a_2 = \beta_2$, the curves have quadratic contact or contact of the second order; and so on. In *n-ic* contact we regard the curves as having $n+1$ consecutive points in common. The transition from n discrete points to n consecutive points requires the notion of an analytic curve connecting the points. Naturally the simplest curve defined by an n-*ic* element is wanted. For $n=1$ it is the line; for $n=2$ we take the circle. For $n=0$ it is the point.

When the second curve is a point on the curve (2), say the base-point, it touches the curve when $a_1=0$. When this happens, the point is a stationary point or cusp. In general there are for a function y of x, points x for which $D_x y=0$. But these need not lie on the selected base-line, so that we do not expect a cusp.

In comparing two curves we took the same parameter ρ. But there is for each curve an appropriate parameter under a given group. Thus for the base-line under translations the proper parameter is the distance, s. To take as parameter a real function of the distance introduces a complication. So under homologies the proper parameter is $\rho=\alpha s+\beta$, and under homographies it is $\rho=\dfrac{\alpha s+\beta}{\gamma s+\delta}$, where in each case the constants are such that ρ is always real. For instance, if we take any real parameter ρ, we may say that any line is

$$x=a_0+a_1\rho$$

and has the differential equation

$$d^2x/d\rho^2=0$$

But we must notice that also

$$d^2x/d\rho^2=\kappa dx/d\rho$$

will be a line when κ is real. For then

$$dx/d\rho=\kappa(x-x_0)$$
$$x-x_0=e^{\kappa(\rho-\rho_0)}$$

It is necessary, then, to discuss the proper parameters.

§ 75. The Homologous Derivative – For the homology

$$y=ax+b$$

we have

$$dy/dx=a$$

(1) $$d^2y/dx^2=0$$

This then characterises the homology itself. If x and y be functions of u, then, denoting differentiations as to u by a superposed dot,

$$\dot{y} = a\dot{x}$$
$$\ddot{y} = a\ddot{x}$$

so that $\ddot{x}/\dot{x}$ is a constant under homologies, applied to x. Let us denote it by (x, u). Then $(x, y) = 0$ when x and y are homologous. But then also $(y, x) = 0$. We look then for an identity which connects (x, y) and (y, x).

If we use the ordinary processes of the calculus, we have

$$\frac{dx}{dy} \cdot \frac{dy}{dx} = 1$$

Differentiate the logarithm of this. Then

$$dx \frac{d^2y}{dx^2} \bigg/ \frac{dy}{dx} + dy \frac{d^2x}{dy^2} \bigg/ \frac{dx}{dy} = 0$$

that is,

$$(y, x)dx + (x, y)dy = 0 \qquad (\mathrm{H}_2)$$

Similarly if x and y, y and z, and therefore z and x be analytically related, from

$$\frac{dx}{dy} \cdot \frac{dy}{dz} \cdot \frac{dz}{dx} = 1$$

we obtain by differentiating the logarithm

$$(y, x)dx + (z, y)dy + (x, z)dz = 0 \qquad (\mathrm{H}_3)$$

This cyclic rule evidently holds in general.

Let us apply it to the planar curve

$$f(x, \bar{x}) = 0$$

The formula (2) gives

$$(\bar{x}, x)dx + (x, \bar{x})d\bar{x} = 0$$

Therefore

$$(x, \bar{x})d\bar{x} = 2\iota d\lambda$$

where λ is real. Apply a homology $x = ay + b$ and use the analogous formula (H_4) for the cycle $x, \bar{x}, y, \bar{y}$. Then

$$(x, \bar{x})d\bar{x} + (\bar{x}, \bar{y})d\bar{y} + (\bar{y}, y)dy + (y, x)dx = 0$$

But here $(x, y)=0$ and $(\bar{x}, \bar{y})=0$. Hence

$$\begin{aligned}(x, \bar{x})d\bar{x} &= -(\bar{y}, y)dy \\ &= (y, \bar{y})d\bar{y}\end{aligned}$$

Hence $d\lambda$ is a constant under homologies, and is the proper real parameter to use for a curve.

The meaning of λ is in this case evident. For

$$dx/d\bar{x} = e^{2\iota\phi}$$
$$d^2x/d\bar{x}^2 = 2\iota e^{2\iota\phi}d\phi/d\bar{x}$$

Whence

$$(x, \bar{x})d\bar{x} = 2\iota d\phi$$

Thus λ is ϕ, the angle of the curve at the point considered. Another form is

$$d\lambda = ds/r$$

We now apply the formula (H_3) to $\bar{x}$, λ, x:

$$(\bar{x}, \lambda)d\lambda + (\lambda, x)dx + (x, \bar{x})d\bar{x} = 0$$

or

$$(\bar{x}, \lambda)d\lambda - (x, \lambda)d\lambda + 2\iota d\lambda = 0$$

or

$$(x, \lambda) - (\bar{x}, \lambda) = 2\iota$$

or

$$(x, \lambda) = \iota + \mathrm{J}$$

where J is real. We call J the fundamental differential invariant (under homologies). Its meaning, again, is simple to find ; for

$$dx = ds e^{\iota\phi}$$
$$dx/d\phi = re^{\iota\phi}$$
$$d^2x/d\phi^2 = (dr/d\phi + \iota r)e^{\iota\phi}$$

whence

$$(x, \phi) = \iota + dr/rd\phi = \iota + dr/ds$$

so that

$$\mathrm{J} = dr/ds = d(\log r)/d\phi$$

Lastly, by the intrinsic equation of a curve (under homologies) we mean

$$\mathrm{J} = f(\lambda)$$

that is,

$$d(\log r)/d\phi = f(\phi)$$

A line has the equation

$$ax + \bar{a}\bar{x} = \rho$$

and is thus characterised among curves by the fact that

$$(x, \bar{x}) \equiv 0$$

If we calculate $(x, \bar{x})$ for any curve given by

$$f(x, \bar{x}) = 0$$

we obtain an expression which, when it vanishes, means that the curve is more than usually like a line. Whereas usually at a point x of a curve we can find a line on x for which $dx/d\bar{x}$ is the same for line and curve (the tangent), we have here $d^2x/d\bar{x}^2$ the same for both.

§ 76. **The Homographic Derivative** – For the homography

$$\begin{aligned} y &= a + b/(c - x) \\ \dot{y} &= b/(c - x)^2 \\ \ddot{y} &= 2b/(c - x)^3 \\ \dddot{y} &= 6b/(c - x)^4 \end{aligned}$$

whence

$$\dddot{y}/\dot{y} = 6/(c - x)^2 = \frac{3}{2}(\ddot{y}/\dot{y})^2$$

Thus y and x are homographically related when

$$\dddot{y}/\dot{y} - \frac{3}{2}(\ddot{y}/\dot{y})^2 \equiv 0 \tag{1}$$

This differential expression is called the Schwarzian derivative; we denote it by $\{y, x\}$. Since the homographic relation is transitive and in particular reversible, there must be an identity between

$$\{x, y\}, \{y, z\} \ . \ . \ . \ \{l, x\}$$

This we shall determine, using power series in preference to differentiations.

Let y be an analytic function of x. To a change ξ in x corresponds a change η in y where

$$\eta = \alpha_1\xi + \alpha_2\xi^2 + \ . \ . \ . = dy + \frac{1}{2\,!}d^2y + \frac{1}{3\,!}d^3y + \ . \ . \ .$$

and

$$\xi = \beta_1\eta + \beta_2\eta^2 + \ . \ . \ . = dx + \frac{1}{2\,!}d^2x + \frac{1}{3\,!}d^3x + \ . \ . \ .$$

Hence
$$\eta = \alpha_1(\beta_1\eta + \beta_2\eta^2 + \ldots) + \alpha_2(\beta_1\eta + \beta_2\eta^2 + \ldots)^2 + \alpha_3(\beta_1\eta + \beta_2\eta^2 + \ldots)^3 + \ldots$$

so that
$$\alpha_1\beta_1 = 1$$
$$\alpha_1\beta_2 + \alpha_2\beta_1^2 = 0$$
$$\alpha_1\beta_3 + 2\alpha_2\beta_1\beta_2 + \alpha_3\beta_1^3 = 0$$
$$\cdot \quad \cdot \quad \cdot \quad \cdot \quad \cdot$$

or expressed as differentials,
$$\frac{dy}{dx} \cdot \frac{dx}{dy} = 1$$
$$\frac{dy}{\xi} \cdot \frac{d^2x}{\eta^2} + \frac{d^2y}{\xi^2}\left(\frac{dx}{\eta}\right)^2 = 0$$
that is,
$$d^2x/dx + d^2y/dy = 0$$
the formula (H_2) of § 75, and
$$\frac{1}{6} \cdot \frac{dy}{\xi} \cdot \frac{d^3x}{\eta^3} + 2\frac{d^2y}{2\xi^2} \cdot \frac{dx}{\eta} \cdot \frac{d^2x}{2y^2} + \frac{1}{6} \cdot \frac{d^3y}{\xi^3} \cdot \left(\frac{dx}{\eta}\right)^3 = 0$$
that is,
$$\frac{d^3x}{dx} + 3\frac{d^2x}{dx} \cdot \frac{d^2y}{dy} + \frac{d^3y}{dy} = 0$$
Hence subtracting $\frac{3}{2} \times$ the square of H_2,
$$d^3x/dx - \frac{3}{2}(d^2x/dx)^2 + d^3y/dy - \frac{3}{2}(d^2x/dx)^2 = 0$$
that is,

(C_2) $$dx^2\{y, x\} + dy^2\{x, y\} = 0$$

Similarly if x, y, z be analytically related each to each, and if ξ, η, ζ be corresponding changes, then
$$\eta = \alpha_1\xi + \alpha_2\xi^2 + \alpha_3\xi^3 + \ldots = dy + d^2y/2\,! + d^3y/3\,! + \ldots$$
$$\zeta = \beta_1\eta + \beta_2\eta^2 + \beta_3\eta^3 + \ldots = dz + d^2z/2\,! + d^3z/3\,! + \ldots$$
$$\xi = \gamma_1\zeta + \gamma_2\zeta^2 + \gamma_3\zeta^3 + \ldots = dx + d^2x/2\,! + d^3x/3\,! + \ldots$$
whence
$$\xi = \alpha_1\beta_1\gamma_1\xi + [\gamma_1(\beta_1\alpha_2 + \beta_2\alpha_1^2) + \gamma_2\alpha_1^2\beta_1^2]\xi^2 + [\gamma_1(\beta_1\alpha_3 + 2\beta_2\alpha_1\alpha_2 + \beta_3\alpha_1^3) + 2\alpha_1\beta_1\gamma_2(\beta_1\alpha_2 + \beta_2\alpha_1^2)]\xi^3 + \ldots$$

so that

$$\alpha_1\beta_1\gamma_1 = 1$$
$$\alpha_2/\alpha_1 + \alpha_1\beta_2/\beta_1 + \alpha_1\beta_1\gamma_2/\gamma_1 = 0$$

or in terms of the differentials,

$$d^2x/dx + d^2y/dy + d^2z/dz = 0$$

the formula (H_3) of § 75, and

$$\alpha_3/\alpha_1 + \alpha_1^2\beta_3/\beta_1 + \alpha_1^2\beta_1^2\gamma_3/\gamma_1$$
$$+ 2\alpha_2\beta_2/\beta_1 + 2\beta_2\gamma_2\alpha_1^2/\gamma_1 + 2\gamma_2\alpha_2\beta_1/\gamma_1 = 0$$

or in terms of differentials,

$$\Sigma d^3x/dx + 3\Sigma d^2xd^2y/dxdy = 0$$

or, subtracting $\frac{3}{2} \times$ the square of (H_3)

$$\Sigma[d^3x/dx - \tfrac{3}{2}(d^2x/dx)^2] = 0$$

whence

$$(C_3) \qquad \{x, y\}dy^2 + \{y, z\}dz^2 + \{z, x\}dx^2 = 0$$

And so in general for any number n of related functions $x, y, z, \ldots l$, we have Cayley's cyclic formula C_n, namely,*

$$(C_n) \qquad \{x, y\}dy^2 + \{y, z\}dz^2 + \ldots + \{l, x\}dx^2 = 0$$

§ 77. **Homographic Invariants of a Curve** – Having this fundamental fact for homographies we proceed as we did for homologies. Taking the case of a planar curve

$$f(x, \bar{x}) = 0$$

we have

$$\{x, \bar{x}\}d\bar{x}^2 + \{\bar{x}, x\}dx^2 = 0$$

so that $\{x, \bar{x}\}dx^2$ is an imaginary, say,

$$\{x, \bar{x}\}d\bar{x}^2 = 2\iota d\lambda^2$$

where $d\lambda^2$ is real. Apply a homography

$$x = (ay + b)/(cy + d)$$

and use Cayley's rule for the cycle $x, \bar{x}, \bar{y}, y$, observing that $\{x, y\} = 0$ and $\{\bar{x}, \bar{y}\} = 0$. Then

$$\{x, \bar{x}\}d\bar{x}^2 = -\{\bar{y}, y\}dy^2$$
$$= \{y, \bar{y}\}d\bar{y}^2$$

* Cayley, *Cambridge Phil. Trans.*, vol. 13, 1880.

Hence $d\lambda^2$ is a constant under homographies, and λ is the proper parameter.

Using now (C_3) for the cycle $\bar{x}$, λ, x,

$$\{\bar{x}, \lambda\}d\lambda^2 + \{\lambda, x\}dx^2 + \{x, \bar{x}\}d\bar{x}^2 = 0$$
$$\{\bar{x}, \lambda\}d\lambda^2 - \{x, \lambda\}d\lambda^2 + 2\iota d\lambda^2 = 0$$
$$\{x, \lambda\} - \{\bar{x}, \lambda\} = 2\iota$$

Hence

$$\{x, \lambda\} = \iota + J$$

where J is real. We call J the fundamental differential invariant of the curve. The intrinsic equation of a curve under homographies is

$$J = f(\lambda)$$

Taking the arc-length s as independent variable we have

$$\dot{x} = e^{\iota\phi}$$
$$\ddot{x} = \iota e^{\iota\phi}c$$

where c is the curvature, $d\phi/ds$. Then

$$\dddot{x} = e^{\iota\phi}(\iota\dot{c} - c^2)$$
$$\{x, s\} = \iota\dot{c} + c^2/2$$
$$\{x, s\} - \{\bar{x}, s\} = 2\iota\dot{c}$$

From the formula (C_3)

$$\{x, \bar{x}\}d\bar{x}^2 = \{x, s\}ds^2 - \{\bar{x}, s\}ds^2$$

Hence

$$2\iota d\lambda^2 = 2\iota\dot{c}ds^2$$
$$d\lambda^2 = dcds$$

As defined $d\lambda$ is real when $\dot{c}$ is positive for increasing s—that is, when the curvature and the arc-length increase together. We notice that at an *apse*—that is, where the curvature is stationary—$d\lambda = 0$. In passing an apse, λ acquires a factor $\pm\iota$, so that the curve where it has apses is mapped on a succession of segments at right angles.

If we wish $d\lambda$ to be always real, we should write

$$\{x, \bar{x}\}d\bar{x}^2 = \pm 2\iota d\lambda^2$$

§ 78. **Special Cases** – J is a number attached to a point of a curve which is constant for all homographies. If then the curve admits homographies into itself, J has the same value at the corresponding points. If the curve admits an infinitesimal homography into itself then J will be a constant. Hence the important curve here is the equiangular spiral, which we saw is built on similar triangles with a fixed vertex (§ 12).

To verify that J is a constant, take the spiral in polar co-ordinates

$$r = \exp(\theta \cot a)$$

so that

$$x = \exp \theta(\iota + \cot a)$$

Hence

$$\{x, \theta\} = -\frac{1}{2}(\iota + \cot a)^2$$

Since from C_3

$$\{x, \bar{x}\} d\bar{x}^2 = \{x, \theta\} d\theta^2 - \{\bar{x}, \theta\} d\theta^2$$

we have

$$2\iota d\lambda^2 = -2\iota \cot a d\theta^2$$

or

$$\lambda = \sqrt{-\cot a} \,.\, d\theta$$

Hence

$$\{x, \lambda\} = -\tan a\{x, \theta\}$$

$$= \frac{1}{2}(\tan a - \cot a) + \iota$$

and

$$J = -\cot 2a$$

In particular when a is $\pm \pi/4$, $J = 0$.

After recent German custom * we may call λ the inversive length of a curve, and J the inversive curvature. The equiangular spiral (or any homograph of it) is then the curve of constant inversive curvature, and the spiral with angle $\pi/4$ (or any homograph) is the curve of zero inversive curvature.

To calculate J in general, we find first $\{\lambda, s\}$. Since

$$\dot{\lambda}^2 = \dot{c}$$

$$2\ddot{\lambda}/\dot{\lambda} = \ddot{c}/\dot{c}$$

$$2[\dddot{\lambda}/\ddot{\lambda} - (\ddot{\lambda}/\dot{\lambda})^2] = \dddot{c}/\dot{c} - (\ddot{c}/\dot{c})^2$$

and

$$\{\lambda, s\} = \frac{1}{2}\dddot{c}/\dot{c} - \frac{5}{8}\ddot{c}(\dot{c}/)^2 \tag{1}$$

* References will be found in B. C. Patterson's paper, "The Differential Invariants of Inversive Geometry," *American Journal*, vol. 50, pp. 553–568.

Since

$$\{x, \lambda\}d\lambda^2 = \{x, s\}ds^2 - \{\lambda, s\}ds^2$$

$$\{x, \lambda\}\dot{c} = \iota c + c^2/2 - \{\lambda, s\}$$

(2) $$\dot{c}\mathrm{J} = c^2/2 - \{\lambda, s\}$$

Exercise 1 – Verify that when $cs = \kappa$, J is a constant.

For the equiangular spiral we have $\phi = \kappa \log s$. To form some idea of other cases, let

(3) $$\phi = \kappa s^n/n$$

where n is not 0. Then

$$c = \kappa s^{n-1}$$

$$\dot{c} = \kappa(n-1)s^{n-2}$$

$$\cdot\ \cdot\ \cdot\ \cdot\ \cdot\ \cdot\ \cdot\ \cdot\ \cdot$$

Here

$$\kappa(n-1)s^n\mathrm{J} = \frac{1}{2}\kappa^2 s^{2n} + \frac{1}{8}(n^2 - 4)$$

or

$$n(n-1)\phi\mathrm{J} = \frac{1}{2}n^2\phi^2 + \frac{1}{8}(n^2 - 4)$$

and

$$d\lambda^2 = \kappa(n-1)s^{n-2}ds^2$$

$$\lambda = 2\sqrt{\kappa(n-1)}s^{n/2}/n$$

(4) $$= 2\sqrt{(n-1)\phi/n}$$

Hence

(5) $$\mathrm{J} = n^2\lambda^2/8(n-1)^2 + (n^2 - 4)/2\lambda^2$$

This, then, is the intrinsic equation of the curve (3), and of all its homographs. The curve is given by

$$x = \int e^{\iota\phi}\phi^{1/n-1}d\phi$$

There are from (3) two cases at first sight, for when s increases, if n be positive, ϕ will increase and the curve will behave spirally; whereas if n be negative there is an asymptotic line $\phi = 0$.

In the former case we expect a limit for x, say,

$$\rho_0 t_0 = \int_0^\infty e^{\iota\phi}\phi^{1/n-1}d\phi$$

Writing $\phi_1 = -\phi$, we then have

$$\rho_0 t_0 = -\int_0^\infty e^{-\iota\phi_1}(-1)^{1/n-1}\phi_1^{1/n-1}d\phi_1$$
$$= (-1)^{1/n}\rho_0/t_0$$

and

$$t_0^2 = (-1)^{1/n}$$

Thus for $n = 2$ we have the Cornu spiral. Here

$$\lambda^2/2 = \phi$$

and the intrinsic equation is

$$\mathrm{J} = \lambda^2/2$$

The singular points have the directions $\pm\sqrt{\iota}$, measured from the tangent at $\lambda = 0$ (fig. 42).

FIG. 42

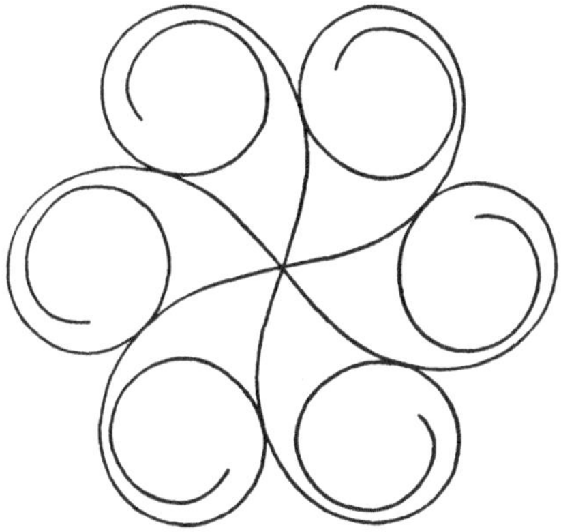

FIG. 43

For $n = 3/2$, we have the spiral

$$8\mathrm{J} = 9\lambda^2 - 7/\lambda^2$$

and the singular points have the directions $\pm 1^{1/3}$ (fig. 43).

Exercise 2 – Apply the recurrence formula

$$e^x x^a/a! = \int_0^x e^x x^a dx/a! + \int_0^x e^x x^{a-1} dx/(a-1)!$$

to test the above conjecture.

In the case when n is negative, and the curve no longer lies in a finite part of the plane, it is proper to invert it. But this is where the intrinsic equation helps. Suppose $n = -2$, so that the intrinsic equation is

$$J = \lambda^2/18$$

Here $\lambda = \sqrt{6\phi}$, $J = \phi/3$. We have then one point for which $J = 0$, one change of sign of λ^2, and two asymptotic points for which $J = \infty$ and $\lambda = \infty$.

§ 79. **Schwarz's Integral** – Let f be a polynomial in x, say

$$f = a_0 + na_1x + \binom{n}{2}a_2x^2 + \quad . \quad . \quad . \quad + a_nx^n$$

A homography

$$x = \frac{ay+b}{cy+d}$$

gives a polynomial in y divided by $(cy+d)^n$. And

$$dx = (ad - bc)dy/(cy+d)^2$$

Hence $dx/f^{2/n}$ is homographic—that is, it acquires under a homography a factor independent of x. Write

$$du = dx/f^{2/n} \tag{1}$$

To regard this as a mapping we must assign the orthogonal sets of paths. Let u describe a line; then du has a given direction. Hence, if the angle of dx is ϕ, and that of $x - x_i$ is θ_i where x_i is a zero of f,

$$\phi - \frac{2}{n}\sum \theta_i = \text{a constant, } \kappa \tag{2}$$

This assigns to any point x, n angles ϕ differing by $2\pi/n$, when n is odd, and m angles differing by π/m when m is even, since θ_i can be increased by 2π. But for a region of the x-plane which does not contain any point x_i, if the angle ϕ is selected for one point it follows by continuity for all points, and the curve for x in this region corresponding to the line for u becomes definite. The curves (2) are inversive.

In the case where the points x_i are all on a line (or a circle), say on the base-line, the curve (2) is for $\kappa = 0$ and $\theta_i = 0$, $\phi = 0$, a segment of a line. As x traverses the base-line from right to left, when it passes x_1 by a small semicircle, say in the upper half-plane, ϕ increases by $2\pi/n$, and is then constant until x passes x_2. Thus the path of u becomes a polygon in the plane of u. Since the total angle turned through by u is 2π, the polygon is closed.

This is Schwarz's mapping of the half-plane on the interior of a polygon, in the case where the roots are all simple, so that the polygon is equiangular. For more general cases see § 80.

We have in (1) a sequence of mappings. Write

$$x = \psi_n u$$

For $n = 1$,

$$du = \kappa dx/(x - x_1)^2$$
$$u - u_1 = \kappa/(x_1 - x)$$

so that ψ_1 is a homography.

Exercise 3 – The curves (2) are here circles on x_1.

For $n = 2$,

$$du = dx/(a_0 + 2a_1 x + a_2 x^2)$$

or, applying a homography to x,

$$du = dx/x$$

Thus ψ_2 is $(ae^u + b)/(ce^u + d)$.

Exercise 4 – The curves $\varphi - (\theta_1 + \theta_2) = \kappa$ are circles on or about x_1, x_2.

For $n = 3$, we have

$$du = dx/(x^3 + 1)^{2/3}$$

and ψ_3 is an elliptic function of u (built on an equilateral network). For mapping with this function, see Adams.*

For $n = 4$,

$$du = dx/(4x^3 - g_2 x - g_3)^{1/2}$$

and ψ_4 is an elliptic function, here the p-function of Weierstrass.

* Oscar S. Adams, "Elliptic Functions applied to Conformal World Maps," *U.S. Department of Commerce Special Publication No.* 112 (Washington, 1925). Geographically, a hemisphere (or other spherical cap) is mapped by ψ_1 into the interior of a circle, by ψ_2 into a strip (Mercator mapping, § 13), by ψ_3 into the interior of an equilateral triangle, and by ψ_4 (in the case when x_i are on a circle) into the interior of a rectangle.

In these cases, and only in these, the function ψ_n is one-valued.

Denoting differentiation as to x by D, the equation (1) gives

$$f = (\mathrm{D}u)^{-n/2}$$

so that

$$\mathrm{D}^{n+1}(\mathrm{D}u)^{-n/2} = 0$$

This is then the differential equation for u. If, replacing n by r we form

$$\mathrm{D}^{r+1}(\mathrm{D}u)^{-r/2}$$

since

$$\mathrm{D}u = 1/f^{2/n}$$

we have

(3) $$\mathrm{D}^{r+1} f^{r/n}$$

This particular operation on f will have the homographic quality, and should be examined.

When $r = 1$, the expression is $\mathrm{D}^2 f^{1/n}$. It vanishes always when f is an nth power. Otherwise it vanishes when

(4) $$nf\ddot{f} - (n-1)f^2 = 0$$

The meaning is that at these points x, f is more nearly an nth power than at others. Thus if

$$f = 1 + n\alpha_1 x + \binom{n}{2}\alpha_2 x^2 + \ldots$$

this and

$$(1 + \alpha_1 x)^n$$

agree to a first approximation, but at the points given by (4) they agree to a second approximation.

If

$$f = \alpha_0 + n\alpha_1 x + \ldots$$

then at the point 0

$$f = \alpha_0, \qquad \dot{f} = n\alpha_1, \qquad \ddot{f} = n(n-1)\alpha_2$$

and (4) is

$$\alpha_0\alpha_2 - \alpha_1{}^2 = 0$$

The points are the Hessian points of the points f.

Exercise 5 – If $\dot{u} = f^{-2/n}$ prove that

$$\{u, x\} = -2(nf\ddot{f} - (n-1)\dot{f}^2)/n^2 f^2$$

Again, there will be points x around which the mapping is the function ψ_2 to a closer approximation than elsewhere. To find them we obtain the differential equation for ψ_2. From (1),

$$\dot{u} = f = a_0 + 2a_1x + a_2x^2$$

so that

$$D^3(Du)^{-1} = 0$$

Thus for ψ_n, for which $Du = f^{-2/n}$, the points in question are given by

$$D^3 f^{2/n} = 0$$

that is by

(5) $$f^2\dddot{f} - 3(n-2)f\dot{f}\ddot{f}/n + 2(n-1)(n-2)\dot{f}^3/n^2 = 0$$

If

$$f = a_0 + na_1x + \binom{n}{2}a_2x^2 + \ldots$$

so that

$$\dot{f} = n[a_1 + (n-1)a_2x + \ldots]$$
$$\ddot{f} = n(n-1)[a_2 + (n-2)a_3x + \ldots]$$
$$\dddot{f} = n(n-1)(n-2)[a_3 + (n-3)a_4x + \ldots]$$

then the constant term in (5) is

$$n(n-1)(n-2)a_0^2a_3 - 3n(n-1)(n-2)a_0a_1a_2 + n(n-1)(n-2)(a_0^2a_3 - 3a_0a_1a_2 + 2a_1^3)$$

and this shows that the covariant in (5) is the one denoted by J, whose symbolic expression is

$$|\alpha\gamma|^2 \, |\beta\gamma| \, (\alpha x)^{n-2}(\beta x)^{n-1}(\gamma x)^{n-3}$$

In general, to calculate $D^{n+1}f^{r/n}$ the method of series with the simplification due to writing $a_0 = 1$, $a_1 = 0$ is convenient.

If the Hessian vanishes, then every point is a Hessian point, and $dx/f^{2/n}$ becomes $dx/(a_0 + a_1x)^2$. Thus f is the nth power of $a_0 + a_1x$. If the covariant J vanishes, then $dx/f^{2/n}$ becomes $dx/(a_0 + 2a_1x + a_2x^2)$. Thus for $n = 2m$, f is the mth power of a quadratic when $J = 0$. And in general if $n = rm$ the covariant which vanishes when f is the mth power of a form of order r is obtained from ψ_r.*

* Compare Hilbert, *Math. Annalen*, vol. 27 (1886), pp. 158–161, where the general form is given.

Exercise 6 – If we develop $\sin(x-a)$ in powers of x, we have

$$\cos a(x - x^3/3! + \ldots) - \cos a(1 - x^2/2! + \ldots)$$

so that $a_0 = -\cos a$, $a_1 = \cos a$, $a_2 = \sin a$, $a_3 = -\cos a$. Prove that

$$a_0 a_2 - a_1^2 = -1$$

and

$$a_0^2 a_3 - 3a_0 a_1 a_2 + 2a_2^3 = -2\cos a$$

The latter vanishes when $a = (2n-1)\pi/2$.

§ 80. **Conformal Mapping** – A line is given by a special homology, say

$$y - y_0 = t(\bar{y} - \bar{y}_0)$$

Thus the homologous derivative $(y, \bar{y})$ vanishes for a line. Hence for any real parameter ρ, $(y, \rho) = (\bar{y}, \rho)$; that is, (y, ρ) is real, when y moves on a line. As ρ describes the base-line, say from right to left, y describes a line until it reaches a point y_1 for which $y - y_1 = c(\rho - \rho_1)^{a_1} + \ldots$, where a is not 1. The value of (y, ρ) is then $(a_1 - 1)/(\rho - \rho_1) + \ldots$ Let ρ describe a small semicircle around ρ, and proceed along the base-line. Then y turns through the angle $\pi - a_1\pi$, and continues along another line.

If then we write

(1) $$(y, \rho) = \Sigma(a_i - 1)/(\rho - \rho_i)$$

we have mapped the base-line into a polygon with interior angles $a_i\pi$.

If for $\rho = \infty$ y is not a singular point, then $y = a/\rho + \ldots$ and $(y, \rho) = -2/\rho + \ldots$ Hence the polygon is closed if

$$\Sigma(a_i - 1) = -2$$

A circle is given by a special homography, say

$$(y - y_0)(\bar{y} - y_0) = \kappa^2$$

Thus the homographic derivative $\{y, \bar{y}\}$ vanishes for a circle. Hence for a real parameter ρ $\{y, \rho\} = \{\bar{y}, \rho\}$; that is, $\{y, \rho\}$ is real when y moves on a circle. Conversely if $\{y, \rho\}$ is real, y will move on arcs of circles, and the change from arc to arc will occur when $y - y_1 = c(\rho - \rho_1)^{a_1} + \ldots$, where a_1 is not 1. That is, when $\{y, \rho\} = (1 - a_1^2)/2(\rho - \rho_1)^2 + \ldots$ Then y turns through the angle $\pi - a_1\pi$, and continues along another arc.

If $\rho=\infty$ is not a singular point, we may write for large ρ

$$y-y_0=1/\rho+b/\rho^2+\ \ldots$$

and calculating $\{y, \rho\}$ we find $-6b^2/\rho^4+\ \ldots$

If then we write

$$\{y, \rho\}=\Sigma \mathrm{R}_i \tag{2}$$

where R_i is a real rational fraction of ρ of degree -4, such that when $\rho=\rho_i$ it is $(1-a_i^2)/2(\rho-\rho_i)^2$, we have mapped the base-line into a closed arc-polygon, and a half-plane into the interior of the polygon.

Thus for three singular points ρ_i the equation

$$\{y, \rho\}=\Sigma(1-a_1^2)(\rho_1-\rho_2)(\rho_1-\rho_3)/2(\rho-\rho_1)^2(\rho-\rho_2)(\rho-\rho_3)$$

maps the half-plane into the interior of a triangle of arcs, with angles $a_i\pi$.*

We assume (as in § 65) a famous theorem of Riemann, that a region with one rim can be mapped isogonally on a half-plane or the interior of a circle. Thus, if x be a point of the base-circle, and y a point of the region, then y is a one-valued analytic function of x, say $y(x)$, and equally x is a one-valued function of y, say $x(y)$. The geodesics of the circle, that is the orthogonal arcs, map into the geodesics of the region.

An analytic function, near $y=0$, may be taken as a convergent series

$$y+\beta_2y^2+\beta_3y^3+\ \ldots$$

Let the limit be ρt. Then

$$\bar{y}+\bar{\beta}_2\bar{y}^2+\ \ldots\ =\rho/t$$

and

$$y+\beta_2y^2+\ \ldots\ =t^2(\bar{y}+\bar{\beta}_2\bar{y}^2+\ \ldots) \tag{3}$$

This is the pencil of geodesics on $y=0$, corresponding to the diameters of the base-circle.

* Of the very extensive literature on conformal mapping we mention Riemann's works; H. A. Schwarz, *Math. Abhandlungen*, vol. 2; F. Klein, *Math. Abhandlungen*, vol. 3; and for the later treatment C. Carathéodory, *Cambridge Tracts*, No. 28, and H. Bateman, *Partial Differential Equations*, chap. 4. Full references are given in L. R. Ford, *Automorphic Functions* (McGraw-Hill Co., New York). An introduction is *Conformal Representation*, by L. Lewent, translated by R. Jones and D. H. Williams (Methuen, London).

In (3) there are no product-terms. That is, $D_y D_{\bar{y}} f = 0$, or Laplace's equation is satisfied.

The osculating circle of (3) at $y = 0$ is

$$y - t^2\bar{y} + \iota\lambda t y\bar{y} = 0$$

when $\beta_2 y^2 - \iota\lambda t y\bar{y} - t^2\bar{\beta}_2\bar{y}^2$ has the factor $y - t^2\bar{y}$; that is, when

$$\iota\lambda t = \beta_2 t^2 - \bar{\beta}_2$$

The osculating circle is then

$$y - t^2\bar{y} + (\beta_2 t^2 - \bar{\beta}_2)y\bar{y} = 0$$

Hence all the circles, for varying t, are on the point

$$y = 1/\beta_2$$

It is useful to know the locus of flexes, and the locus of apses, for the pencil (3). For this pencil is the paths of flow from a source at $y = 0$ to the rim (acting as sink).

For a flex on a curve, $(y, \bar{y})$ is 0 (§ 75). Hence, if we are mapping a curve x on the curve y,

$$(x, y)dy = (x, \bar{y})d\bar{y}$$

If, then, we are mapping the radii $x = \rho t_0$ on the region y, so that dx/x is real,

$$(x, y)x dy/dx \text{ is real.}$$

This gives the flex-locus for the paths of the region.

For an apse on a curve $\{y, \bar{y}\} = 0$ (§ 76).

Hence

$$\{x, y\}dy^2 \text{ is real.}$$

In the same mapping, we have then

$$\{x, y\}x^2(dy/dx)^2 \text{ is real.}$$

This gives the apse-locus for the paths of the region.

Exercise 7 – A half-plane x is mapped on the interior y of a rhombus by

$$dy = dx(x^2 - 1)^{(\lambda-1)/2}(x^2 + 1)^{-(\lambda-1)/2}$$

Hence determine the flex-locus for the flow from the centre of the rhombus to its edges.

Exercise 8 – Determine the flow from one vertex of a quadrangle to the opposite vertex, the edges being levels; and show that the flex-locus is the map of a circle in the half-plane.

PART II

CHAPTER XII

THE LINE AND THE CIRCLE

§ 81. **Map-equations of a Line** – The translation

$$x = y + \iota\rho_0$$

sends the base-line $y = \rho$ into any parallel line. The rotation

$$x - \rho_0 = t(y - \rho_0)$$

sends the base-line into any line cutting the base-line. Thus all lines may be written either as

$$x = \rho + \iota\rho_0$$

or

$$x - \rho_0 = t(\rho - \rho_0)$$

The first may be derived from the second by letting t approach 1, and ρ_0 approach ∞.

The general homology

$$x = ay + b$$

gives, for $y = \rho$, the line in the more general form

$$x = a\rho + b$$

where a and b are given.

At first sight the four parameters, two for a and two for b, are too many, since two parameters determine a line. But in equation (1) of § 80

$$\rho = \alpha\rho' + \beta$$

we have two disposable parameters α and β. The number of parameters in

$$x = a\rho + b$$

is then $4 - 2$, as it should be. We may, for instance, include the real factor of a in ρ, so that any line is

$$x = t\rho + b$$

where t and b are given. We can still write $\rho + \beta$ for ρ, so that we can replace b by $b + t\beta$, and β can be chosen so that this is a real, when the line cuts the base-line ; or an imaginary, when the line cuts the vertical axis ; or zero, when it cuts both.

Let x be the centre of any stretch which sends a into b. Then

$$x - b = \rho(x - a)$$

so that

$$x = (b - \rho a)/(1 - \rho)$$

For variable ρ this is again the equation of a line, the join of a, b.

Let x be the centre of a rotation which sends a into b. Then

$$x - b = t(x - a)$$

so that

$$x = (b - ta)/(1 - t)$$

For variable t this is the equation of a line, the axis or middle line of a, b. In particular, for $a = 0$,

$$x = b/(1 - t)$$

where b is the image of the base-point.

Such parametric equations are called *map-equations*. The base-line or the base-circle is mapped on to the new line or circle. The last equation is noticeable, because it maps a circle on to a line. The point $t = 1$ gives the point $x = \infty$.

Two points divide their join into two segments, a finite and an infinite segment. It is easy to express either by a map-equation. For

$$2x = t + 1/t$$

or

$$x = \cos\theta$$

is the finite segment with end-points ± 1. And

$$2x = \rho + 1/\rho$$

or

$$x = \cosh\rho$$

is the infinite segment with the same end-points. Applying a homology we have for any finite segment

$$x = a + b(t + 1/t)$$

and for any infinite segment

$$x = a + b(\rho + 1/\rho)$$

§ 82. **Self-conjugate Equations of a Line** – If we write the conjugate of any of these equations and eliminate the parameter we have an equation between x and $\bar{x}$, which is the self-conjugate equation of the line.

Thus from

$$x = b/(1-t)$$
$$\bar{x} = \bar{b}(1-1/t)$$

we have

(1) $$x/b + \bar{x}/\bar{b} = 1$$

the standard equation of a line.

If the line be on the base-point, we may write

$$x = t^2\bar{x}$$

Here t^2 is the clinant of the line, the square of either direction.

All cases are included in the general self-conjugate linear equation

$$ax + \bar{a}\bar{x} + \rho = 0$$

The coefficients here enter homogeneously.

The reflexion in a line given by a self-conjugate equation is obtained by writing y for x, or $\bar{y}$ for $\bar{x}$. Thus we regard a line and generally a curve as bound up with a transformation, the curve being the locus of fixed points. The essential is the ease of transition from the curve to the transformation, and conversely.

Exercise 1 – The directed distance from a point y to the line

$$x/b + \bar{x}/\bar{b} = 1$$

is

$$b(1 - y/b - \bar{y}/\bar{b})/2$$

Other ways of writing a self-conjugate equation of a line are:

(2) In terms of the clinant τ and a point x_1 on the line,

$$x - x_1 = \tau(\bar{x} - \bar{x}_1)$$

If x_1 be where the line meets the base-line, so that $x_1 = \rho$, then

(2) $$x - \rho = \tau(\bar{x} - \rho)$$

Here ρ is the intercept of the line. When $\rho = \infty$, τ becomes 1, but lim $\rho(1-\tau)$ is finite. The limit of the direction of $(1-\tau)$ is ι, so that lim $\rho(1-\tau)$ is an imaginary.

(3) The join of two points x_1, x_2 is

$$\begin{vmatrix} x & x_1 & x_2 \\ \bar{x} & \bar{x}_1 & \bar{x}_2 \\ 1 & 1 & 1 \end{vmatrix} = 0 \tag{3}$$

This expresses that three points on a line form a triangle negatively similar to itself.

For, from the general Euclidean operation,

$$x = a\bar{y} + b$$

the triangles x_i and y_i are negatively similar when

$$| x_i \quad \bar{y}_i \quad 1 | = 0$$

They are positively similar when $x_i = ay_i + b$; that is, when

$$| x_i \quad y_i \quad 1 | = 0$$

(4) The axis of two points x_1, x_2 is

$$\begin{vmatrix} x & x_1 & x_2 \\ \bar{x} & \bar{x}_2 & \bar{x}_1 \\ 1 & 1 & 1 \end{vmatrix} = 0 \tag{4}$$

This expresses that the isosceles triangle x, x_1, x_2 is negatively similar to the triangle x, x_2, x_1. The four determinants of the matrix

$$\begin{vmatrix} a & b & x & y \\ \bar{a} & \bar{b} & \bar{y} & \bar{x} \\ 1 & 1 & 1 & 1 \end{vmatrix}$$

vanish when x, y are images in the join of a, b. The four determinants of the matrix

$$\begin{vmatrix} a & b & x & y \\ \bar{b} & \bar{a} & \bar{y} & \bar{x} \\ 1 & 1 & 1 & 1 \end{vmatrix}$$

vanish when x and y, a and b, have a common axis.

(5) For a line which cuts the base-circle at the points t_1, t_2, we have in the standard form

$$b = t_1 + t_2$$
$$\bar{b} = 1/t_1 + 1/t_2$$

Thus the standard form becomes the equation of a secant,

$$x + \bar{x}t_1t_2 = t_1 + t_2 \tag{5}$$

Any given number of lines may be taken in this form, since the radius of the base-circle—that is, our unit of length—is as large as we please.

Exercise 2 – The lines

$$x - \varrho_1 = \tau(\bar{x} - \varrho_1)$$
$$x - \varrho_2 = -\tau(\bar{x} - \varrho_2)$$

are perpendicular.

Exercise 3 – The joins of b, c and of a, d are perpendicular when

$$(b - c)/(a - d) = \varrho\iota$$

Exercise 4 – From the identity

$$\frac{b-c}{a-d} + \frac{c-a}{b-d} + \frac{a-b}{c-d} + \frac{(b-c)(c-a)(a-b)}{(a-d)(b-d)(c-d)}$$

deduce that the perpendiculars or altitude-lines of a triangle a, b, c meet at a point d.

§ 83. **The Base-circle** – The simplest map-equation of the base-circle is

$$x = t$$

The equation of the line cutting the circle at t_1 and t_2 is

$$x + \bar{x}t_1t_2 = t_1 + t_2$$

Writing this in the form

$$(x - t_1)(\bar{x} - 1/t_2) = x\bar{x} - 1$$

it says that the product of the lengths of the segments of the secant, measured from x to the circle, is the constant $x\bar{x} - 1$. This constant is the power of the point and circle. It is the constant of the inversion with centre x which sends the circle into itself. The pairing or involution thus set up on the circle is elliptic, hyperbolic, or parabolic according as x is inside, outside, or on the circle. In the parabolic case, where x is, say t_1, t_2 is arbitrary, but we assign each point t_2 to the clinant of its join to t_1.

Exercise 5 – If the secant t_1, t_2 be parallel to t_4, t_5, and t_2, t_3 be parallel to t_5, t_6, then t_3, t_4 is parallel to t_6, t_1. This is a form of Pascal's theorem.

Writing again the equation of the secant in the form

$$(x-t_1)/(1-\bar{x}t_1)=t_2$$

or

$$(x-t_1)/(1/\bar{x}-t_1)=t_2\bar{x}$$

we see that the ratio of the distances of any point t_1 of the circle from the given points x and $1/\bar{x}$ is the constant $\bar{x}$. This is a theorem of Apollonius.

As a limiting case of the secant, the tangent-line or simply the line of the circle at t is

(1) $$x+\bar{x}t^2=2t$$

The roots of this quadratic being t_1 and t_2,

$$t_1+t_2=2/\bar{x}$$

Thus the tangents at t_1 and t_2 meet at

(2) $$x=2t_1t_2/(t_1+t_2)$$

For varying t_2, this is a map-equation of the tangent at t_1.

For two points t_i on the circle we have the equation

(3) $$x+\bar{y}t_1t_2=t_1+t_2$$

expressing that x and y are images in a secant. It must be noticed that

(4) $$x+\bar{y}ab=a+b$$

is the antilogy which sends $x=a$ into $y=1/\bar{a}$ and $x=b$ into $y=1/\bar{b}$. This antilogy, an operation defined by a circle and two points, becomes a reflexion in two cases, when both points are on the circle, and when the points are images in the circle. The antilogy is general, for $x+\bar{y}\alpha=\beta$ has in common with $x\bar{y}=1$ the two points given by $x^2+\alpha=\beta x$.

Exercise 6 – Equation (4) expresses that x, a, b is negatively similar to y, $1/\bar{a}$, $1/\bar{b}$.

In particular when $a=b$ we have the tangent antilogy

(5) $$x+\bar{y}a^2=2a$$

with fixed point given by $x+\bar{x}a^2=2a$.

Thus we see what happens to the tangents from a point x to a circle when x is inside. The two values of a are an image-pair; instead of the tangents we have an antilogous form (5), giving two reciprocal antilogies, one of which sends a into $1/\bar{a}$, and the other sends $1/\bar{a}$ into a.

And in general for the tangents from a point to an algebraic curve we have lines of the curve and tangent antilogous forms.

§ 84. **Envelopes** – We have said that the lines t_1, t_2 of the base-circle meet at the point

$$x = 2t_1t_2/(t_1+t_2)$$

This is any point outside the circle.

Generally, when $x = f(t, \tau)$ we have for given t a curve, for then varying t a system of curves; and for given τ a curve, and for then varying τ a second system of curves. We regard x as an intersection of two curves, one of each system; to obtain the envelope we make the two curves touch. When t varies, the direction is given by dx/dt; since $dt/t = \iota d\theta$ the direction is that of $dx/d\theta$, or of $\iota t dx/dt$.

So when τ varies the direction of the other curve is that of $\iota\tau dx/d\tau$.

Hence for the envelope we are to have

(1) $$t\frac{\partial f}{\partial t}\bigg/\tau\frac{\partial f}{\partial \tau} = \text{real}.$$

Exercise 7 – When $x = t + \tau - t\tau$, the envelope is the curve $x = 2t - t^2$, and the point $x = 1$.

Exercise 8 – Circles have their centres on a circle and touch a diameter of it. Find their envelopes.

When x is a function of two reals, $x = f(r, \rho)$, the rule of envelopes is simply that

(2) $$\frac{\partial f}{\partial r}\bigg/\frac{\partial f}{\partial \rho} \text{ is real.}$$

And when $x = f(\rho, t)$, the rule is that

(3) $$\iota t\frac{\partial f}{\partial t}\bigg/\frac{\partial f}{\partial \rho} \text{ is real.}$$

§ 85. **Map-equations of a Circle** – The simplest map-equation of any circle is

(1) $$x = a + bt$$

where a is the centre and $|b|$ the radius.

We may regard this as the result of a homology $x = a + by$ applied to the base-circle.

The group of homologies which send the base-circle into a given circle is then

$$x = a + r\tau y$$

Their centres are given by $x = a/(1 - r\tau)$.

This is a circle covariant under homologies called the circle of similitude. The group contains two stretches, namely, when $\tau = \pm 1$. Their centres are the centres of similitude. They divide the centres of the circles in the ratio of the radii (positively or negatively).

Exercise 9 – They are the centres of the two inversions which interchange the circles.

Exercise 10 – If a centre of similitude is on one circle it is on both, and the circles touch.

Exercise 11 – The six centres of similitude of two out of three circles are the points of a four-line, and the centres of the circles form the diagonal triangle.

If we take the two circles as

$$x - 2t + \bar{x}t^2 = 0$$
$$x - \mu - 2r\tau + (\bar{x} - \mu)\tau^2 = 0$$

Then for common lines we have $t^2 = \tau^2$—that is, $t = \pm\tau$. With $t = \tau$, we have also

$$\mu - 2(1 - r)t + \mu t^2 = 0$$

or

$$t + 1/t = 2(1 - r)/\mu$$

If then $| 1 - r | > \mu$ we have not common lines, but a common antilogous form. For $t = -\tau$ we change the sign of r—that is, the way of description of one circle.

The general one-to-one map-equation of a circle is

$$x = (\alpha_0 + \alpha_1 z)/(\beta_0 + \beta_1 z)$$

where z is a real ρ or a turn t as is convenient. To find the centre, that is the image of ∞, we note that $x = \infty$ when $z = -\beta_0/\beta_1$.

The image of this in the base-line (or base-circle) gives then the centre.

§ 86. Self-conjugate Equation of a Circle – The bilinear curve is homogeneously written

(1) $$f \equiv \rho x\bar{x} - \bar{a}x - a\bar{x} + \sigma = 0$$

In terms of the centre c it is

(2) $$(x-c)(\bar{x}-\bar{c}) + \kappa = 0$$

It is a circle when κ is negative, a double point when $\kappa = 0$, an extra pair, images as to the plane considered, when κ is positive. We may denote it by (c, r^2) in the first case, and by $(c, -\zeta^2)$ in the third case, $\pm\zeta$ being the ordinates to the plane at c.

The discriminant of (1) is $\rho\sigma - a\bar{a}$; the curve is a circle when this is negative.

Replacing $\bar{x}$ by $\bar{y}$ (or x by y) we have from (1) the general inversion $\rho x\bar{y} - \bar{a}x - a\bar{y} + \sigma = 0$, this being hyperbolic, parabolic, or elliptic as its discriminant $\rho\sigma - a\bar{a}$ is negative, 0, or positive.

Exercise 12 – Under an inversion $(0, \kappa)$ the circles (c, r^2) and (c_1, r_1^2) are interchanged, where

$$c_1/c = \bar{c}_1/\bar{c} = \kappa/(c\bar{c} - r^2) = (c_1\bar{c}_1 - r_1^2)/\kappa$$

Two circles (c_i, r_i^2) are orthogonal when

(3) $$r_1^2 + r_2^2 - (c_1 - c_2)(\bar{c}_1 - \bar{c}_2) = 0$$

For the homogeneous forms (1), (3) becomes, since $c_i = -\bar{a}_i/\rho_i$ and $r_i^2 = (a_i\bar{a}_i - \rho_i\rho_i')/\rho_i^2$,

$$\rho_1\sigma_2 - a_1\bar{a}_2 - \bar{a}_1 a_2 + \rho_2\sigma_1 = 0$$

The expression is the bilinear invariant (or polar) of the two curves. For coincident curves it is twice the discriminant.

Denoting it by f_{12}, then $f_{12}/\sqrt{f_{11}}\sqrt{f_{22}}$ is the fundamental constant of two circles, under homographies. When they meet it is from (3) the cosine of an angle of intersection; when they touch it is ± 1; when they do not meet it is $\cosh\lambda$, where λ is a hyperbolic distance.

Exercise 13 – When a given circle touches an extra pair at a given point, the locus of the latter is two lines, inclined to the plane considered at 45°.

§ 87. **The n-line** – Let us write the equation of a line in the form

$$\bar{x} = -(x - x_1)\bar{x}_1/x_1$$

where x_1 is the image of the base-point.

Denoting the reciprocal of the clinant by t_1, this is

(1) $$\bar{x} = t_1(x - x_1)$$

where

(2) $$t_1 = -\bar{x}_1/x_1$$

We have then for two lines

$$\bar{x} = t_i(x - x_i)$$

whence the intersection is

(3) $$x_{12} = x_1 t_1/(t_1 - t_2) + x_2 t_2/(t_2 - t_1)$$

For three lines we have three such intersections. All are included in

(4) $$x = \sum^{3} x_1 t_1(t_1 - \tau)/(t_1 - t_2)(t_1 - t_3)$$

since this when $\tau = t_3$ is x_{12}, and so on.

Now (4) is of the form

$$x = c_0 - c_1\tau$$

It is then the circumcircle of the three-line. Thus the circumcentre, x_{123}, is

(5) $$x_{123} = c_0 = \sum^{3} x_1 t_1^2/(t_1 - t_2)(t_1 - t_3)$$

and the radius is $|\, c_1 \,|$ where

$$c_1 = \sum^{3} x_1 t_1/(t_1 - t_2)(t_1 - t_3)$$

For four lines there are four circumcentres. All are included in

(6) $$x = \sum^{4} x_1 t_1^2(t_1 - \tau)/(t_1 - t_2)(t_1 - t_3)(t_1 - t_4)$$

since when $\tau = t_4$ this is x_{123}, and so on.

Hence the four circumcentres lie on a circle whose centre is

(7) $$x_{1234} = \sum^{4} x_1 t_1^3/(t_1 - t_2)(t_1 - t_3)(t_1 - t_4)$$

and whose radius is $| c_1 |$ where

$$c_1 = \sum^4 x_1 t_1^2/(t_1 - t_2)(t_1 - t_3)(t_1 - t_4)$$

And the argument can be continued indefinitely. Thus we have for n lines no two of which are parallel a covariant point c_0 (under homologies) which is the natural base-point, and a covariant circle which is the natural base-circle. We call this point the centric point, or simply the centric, and this circle the centric circle.

Let us repeat the argument more generally. Write $f(t)$ for

$$(t - t_1)(t - t_2) \; . \; . \; . \; (t - t_n)$$

Then $f'(t_1)$ is

$$(t_1 - t_2)(t_1 - t_3) \; . \; . \; . \; (t_1 - t_n)$$

where n is the number of lines considered.

We have for a three-line the circumcircle

$$x = \sum^3 x_1 t_1 (t_1 - \tau)/f'(t_1) \tag{8}$$

For a four-line we have the expression

$$x = \sum^4 x_1 t_1 (t_1 - \tau_1)(t_1 - \tau_2)/f'(t_1) \tag{9}$$

This when τ_2 is t_4 becomes (8).

It thus includes the four circumcircles of the lines taken three at a time.

Let us write (9) as

$$x = c_0 - c_1(\tau_1 + \tau_2) + c_2 \tau_1 \tau_2. \tag{10}$$

Then

$$c_0 = \sum^4 x_1 t_1^3/f'(t_1)$$

$$c_1 = \sum^4 x_1 t_1^2/f'(t_1)$$

$$c_2 = \sum^4 x_1 t_1/f'(t_1)$$

Hence we have

$$\bar{c}_1 = s_4 c_2$$

where

$$s_4 = t_1 t_2 t_3 t_4$$

and

$$\bar{c}_2 = s_4 c_1$$

For a canonical form of (10) we may then take $c_0=0$, $c_1=-1$, and

$$x=\tau_1+\tau_2-\tau_1\tau_2.$$

The envelope of this (§ 84) will be a curve or curves touched by the four circumcircles. This envelope is:

(1) the curve $x=2\tau-\tau^2$

(2) the point $x=1$.

In particular then the circumcircles of 3/4 lines meet at a point (fig. 44). And this point is on the centric circle.

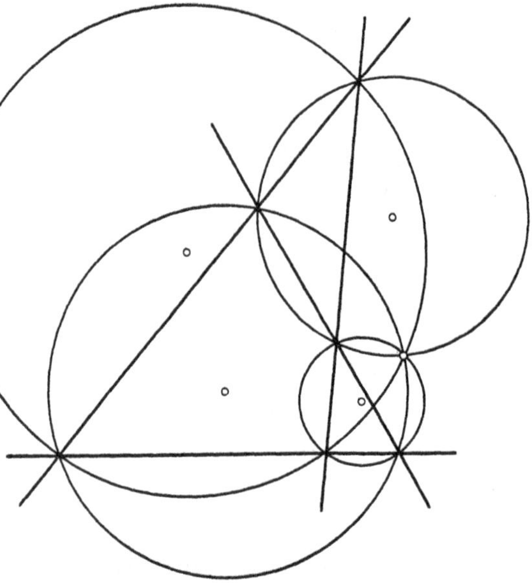

FIG. 44

So for a five-line we write the expression

$$x=\sum^5 x_1t_1(t_1-\tau_1)(t_1-\tau_2)(t_1-\tau_3)/f'(t_1) \tag{11}$$

This when $\tau_3=t_5$ is (9).

We infer that there is intimately connected with the five-line (in Euclidean geometry) a curve

$$x=\sum^5 x_1t_1(t_1-\tau)^3/f'(t_1)$$

or

$$x=c_0-3c_1\tau+3c_2\tau^2-c_3\tau^3, \tag{12}$$

where

$$c_0 = \sum^5 x_1 t_1^4 / f'(t_1)$$
$$c_1 = \sum^5 x_1 t_1^3 / f'(t_1)$$
$$c_2 = \sum^5 x_1 t_1^2 / f'(t_1)$$
$$c_3 = \sum^5 x_1 t_1 / f'(t_1)$$
$$\bar{c}_1 = -s_5 c_3$$
$$\bar{c}_2 = -s_5 c_2$$
$$s_5 = t_1 t_2 t_3 t_4 t_5$$

and in general with an n-line is intimately connected a curve

$$x = \sum^n x_1 t_1 (t_1 - \tau)^{n-2} / f'(t_1)$$
$$= c_0 - n c_1 \tau + \ . \ . \ .$$

where

$$c_i = \sum^n x_1 t_1^{n-1-i} / f'(t_1)$$

and

(13) $$\bar{c}_i = (-)^n s_n c_{n-1-i}$$

We shall consider these curves in Chapter XXI. Here we notice that while c_0 is a covariant point, c_1, c_2 . . . are constants under translations. They form with the n clinants a system of constants for the n-line.

The five centric circles for 4/5 lines are included in

$$x = \sum^5 x_1 t_1^2 (t_1 - \tau_1)(t_1 - \tau_2) / f'(t_1)$$

or

$$x = c_0 - c_1(\tau_1 + \tau_2) + c_2 \tau_1 \tau_2$$

The envelope of this will be touched by the five circles.

By the rule of envelopes

$$\tau_1(c_1 - c_2\tau_1)/(\bar{c}_1\tau_2 - \bar{c}_2) = \tau_2(c_1 - c_2\tau_1)/(\bar{c}_1\tau_1 - \bar{c}_2)$$

so that either $\tau_1 = \tau_2$ or

$$\bar{c}_1\{c_1(\tau_1 + \tau_2) - c_2\tau_1\tau_2\} = c_1\bar{c}_2$$

Thus the envelope is the curve

$$x = c_0 - 2c_1\tau + c_2\tau^2$$

and the point

(14) $$x = c_0 - c_1\bar{c}_2/\bar{c}_1$$

That is, the five centric circles meet at a point. The argument subsists unchanged for $(n-1)/n$ lines.

Thus the n centric circles, for each $(n-1)$-line in an n-line, all meet at the point (14).*

Exercise 14 – The value of x in (14) makes $c_0 - c_1(t+\tau) + c_2 t\tau - x$ self-conjugate; then for any t there is a turn τ. This could be used instead of finding the envelope, in proving the theorem.

§ 88. **Stretches** – The argument of the last section can be applied in other cases. Take the case of stretches. It is convenient to think of a map, say of an island. Under a stretch we get a second positively similar map, the north to south line agreeing. To consider several such maps, write

$$\mu_i x_i = x + a_i$$

where μ_i is real. As x travels over the given map, x_i will travel over say M_i. The fixed point for M_1M_2 is

$$x_{12} = a_1/(\mu_1 - \mu_2) + a_2/(\mu_2 - \mu_1)$$

We write

$$x = \sum^3 a_1(\mu_1 - \mu)/(\mu_1 - \mu_2)(\mu_1 - \mu_3)$$

This is a line, on which lie x_{23}, x_{31}, x_{12}. On the line is the point

$$x_{123} = \sum^3 a_1\mu_1/(\mu_1 - \mu_2)(\mu_1 - \mu_3)$$

for which

$$x_{123} - x_{23} = \mu_1 \sum a_1/(\mu_1 - \mu_2)(\mu_1 - \mu_3)$$

Call this the centric point for the three maps. For four maps we write

$$x = \sum^4 a_1\mu_1(\mu_1 - \mu)/f'(\mu_1)$$

This when $\mu = \mu_4$ is x_{123}.

Thus the centric points for 3/4 maps are on a line. And so on.

* This theorem is attributed to de Longchamps, *Nouvelles Correspondances de Math.*, vol. iii. (1887), by Prof. Coolidge, *Treatise on the Circle and the Sphere*, Clarendon Press, 1916. This is a valuable work of reference.

Exercise 15 – The equation

$$x/t + \bar{x}t - 2\rho = 0$$

is that of a directed line. If we define the distance of a point from n such lines as the sum of the distances, then the point at zero distance is on the line

$$x\bar{s}_1 + \bar{x}s_1 - 2\Sigma\rho = 0$$

Call this the zero-line.

Prove that the zero-line of two directed lines is the bisector of the angle between them, and that the zero-lines of m directed lines and n directed lines meet on the zero-line of the $m+n$ lines. Thus we have on the zero-line of n lines $2^{n-1} - 1$ points. The figure can be constructed by paper-folding.

Exercise 16 – State the similar theorem for planes in a space.

It remains to relate a set of points of the plane to a set of lines.

Exercise 17 – There is one point x which is the centroid of its images in n given lines.

This puts the n-lines of a plane in a one-to-one correspondence with the n-points. So that a comitant (that is an invariant or covariant) of an n-line, under the Euclidean group, is a comitant of the related n-point, and conversely.

Exercise 18 – Four directed lines

$$x/t_i + \bar{x}t_i = 2r_i$$

touch a circle if

$$\sum^{4} r_i t_i / f'(t_i) = 0$$

where $f(t) = (t - t_1) \; . \; . \; . \; (t - t_4)$.

Five directed lines touch a circle if

$$\sum^{5} r_i t_i / f'(t_i) = 0$$

and

$$\sum^{5} r_i t_i^2 / f'(t_i) = 0$$

And so on.

CHAPTER XIII

REGULAR POLYGONS

§ 89. **The Regular Pentagon** – A rotation $y = tx$, effected n times, sends x into $t^n x$. This is x itself when $t^n = 1$. In this case the points of the plane are arranged in regular polygons. To construct a regular polygon we have then to solve the equation $t^n = 1$. We give here a few examples.

For the pentagon $t^5 - 1 = 0$, or removing the factor $t - 1$,

$$t^2 + 1/t^2 + t + 1/t + 1 = 0$$

We write

$$\mu_1 = t + 1/t$$
$$\mu_2 = t^2 + 1/t^2$$

so that $m_i = \mu_i/2$ is the mean of t^i and $1/t^i$. The construction is effected by finding m_i and erecting verticals to cut the base-circle.

We have

$$(1) \qquad \mu^2 + \mu - 1 = 0$$

Consider the circle on the two points μ_1 and μ_2, with centre on the base-line. It has the centre $-1/2$, and meets the vertical axis at $\pm\iota$. If we polarise (1) we have

$$\mu\mu' + \frac{1}{2}(\mu + \mu') - 1 = 0$$

This gives the pairs of reals which are inverse as to the sought circle. The lines from ι to μ_1 and μ_2 are the bisectors of the angle $\mu\iota\mu'$. Thus taking $\mu = 0$, $\mu' = 2$ the construction for μ_1, μ_2 is to bisect the angle $0, \iota, 2$; or for m_1, m_2 is to bisect the angle $0, \iota/2, 1$ (fig. 45).

Analytically, instead of ‘ completing the square ’ we write (1) in the form

$$\left(\frac{1 + \iota\mu}{1 - \iota\mu}\right)^2 = \frac{1 + 2\iota}{1 - 2\iota}$$

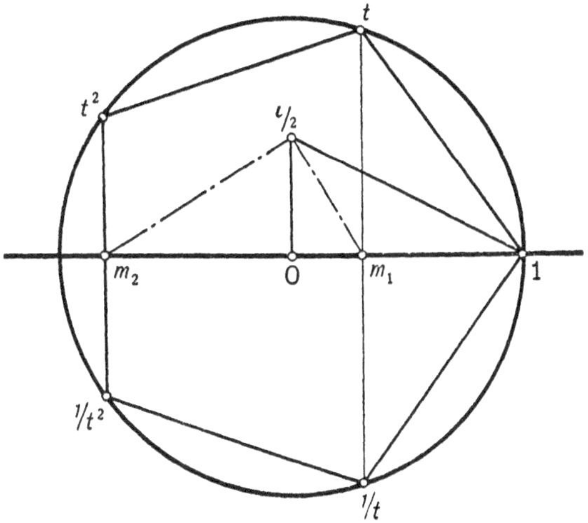

FIG. 45

which expresses that the angle 0, ι, 2 is bisected by the lines from ι to μ.

Exercise 1 – Gauss showed in his *Disquisitiones Arithmeticæ* (1801) that constructions are possible by bisections alone where n is a prime number of the form 2.2^r+1, of which the next case is $n=17$. Here we have eight points $\mu_i = t^i + 1/t^i$ where $\sum^8 \mu_i + 1 = 0$. If we write

$$\nu_1 = \mu_1 + \mu_8 + \mu_4 + \mu_2$$
$$\nu_3 = \mu_3 + \mu_7 + \mu_5 + \mu_6$$

and also

$$\xi_1 = \mu_1 + \mu_4$$
$$\xi_2 = \mu_8 + \mu_2$$
$$\xi_3 = \mu_3 + \mu_5$$
$$\xi_6 = \mu_7 + \mu_6$$

the construction may be effected by the three steps :

(1) Bisect angle 0, $\pm\iota/4$, 1 to find $\nu_1/8$, $\nu_3/8$.
(2) Bisect angles 0, $\pm\iota/4$, $\nu_3/8$ and 0, $\pm\iota/4$, $\nu_1/8$ to find $\xi_1/4$, $\xi_2/4$ and $\xi_3/4$, $\xi_6/4$.
(3) Bisect angle 0, $\pm\iota\sqrt{\xi_2/4}$, ξ_2/ξ_3 to find $m_3 = \mu_3/2$, $m_5 = \mu_5/2$.

These steps are shown in fig. 46. In the third step $q = \xi_2/\xi_3$ is constructed by the proportion

$$\frac{q}{-\iota\xi_6/4} = \frac{-1}{\iota\xi_1/4}$$

and $h = \pm\iota\sqrt{\xi_2/4}$ is found by cutting the ι-axis with a circle on $\xi_2/4$, 1 as diameter.

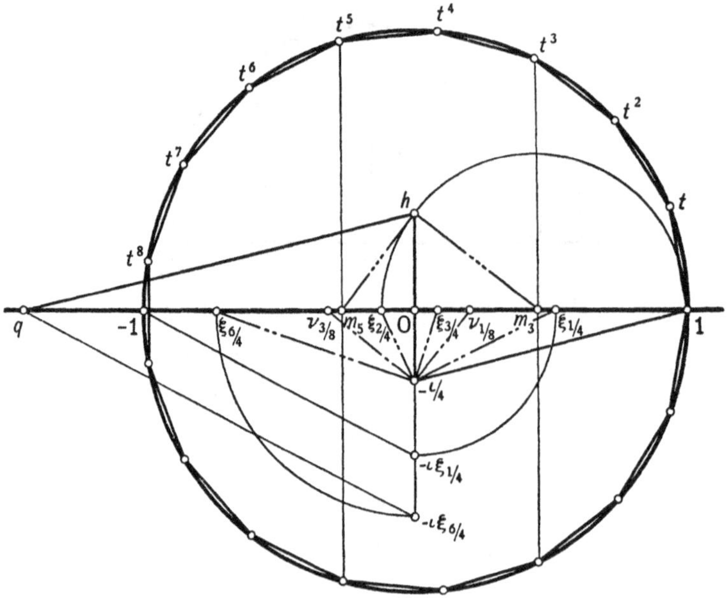

FIG. 46

This form of the construction is adapted from H. W. Richmond, *Quarterly Journal of Mathematics*, vol. 26 (1893), pp. 206, 207. *Cf.* H. P. Hudson's *Ruler and Compasses* (London, 1916), p. 34. We refer also to Klein's *Famous Problems in Elementary Geometry*, translated by Beman and Smith (Boston, 1897), and for a discussion of the algebra and a proof of Gauss's Theorem, to Weber's *Algebra*.

§ 90. **The Regular Heptagon** – Here

$$t^3 + 1/t^3 + t^2 + 1/t^2 + t + 1/t + 1 = 0$$

Let

$$\mu_1 = t + 1/t = t^6 + 1/t^6$$
$$\mu_2 = t^2 + 1/t^2 = t^5 + 1/t^5$$
$$\mu_3 = t^3 + 1/t^3 = t^4 + 1/t^4$$

Then

$$\mu_1 + \mu_2 + \mu_3 = -1$$
$$\mu_2\mu_3 = \mu_1 + \mu_2, \quad \mu_3\mu_1 = \mu_2 + \mu_3, \quad \mu_1\mu_2 = \mu_3 + \mu_1$$
$$\mu_2\mu_3 + \mu_3\mu_1 + \mu_1\mu_2 = -2$$
$$\mu_1\mu_2\mu_3 = \mu_1{}^2 + \mu_1\mu_2 = 1$$

The μ_i are then given by the cubic

$$3(\mu^3 + \mu^2 - 2\mu - 1)$$

of which the Hessian is

$$-7(h^2+h+1)$$

Thus the Hessian points are ω and ω^2. We may now apply § 42; but it is better in this case of three real roots to use lines only. We seek the coincidence points of the involution along the axis of reals

$$3pqr+qr+rp+pq-2(p+q+r)-3=0$$

whose neutral pair is ω, ω^2. Putting these in evidence, the involution is

$$(p-\omega)(q-\omega)(r-\omega)=\kappa(p-\omega^2)(q-\omega^2)(r-\omega^2)$$

and for the coincidence points

$$(\mu-\omega)^3=\kappa(\mu-\omega^2)^3$$

Hence

$$(\mu-\omega)^3/(p-\omega)(q-\omega)(r-\omega)$$

is real; that is, the product of the directions from ω to p, q, r is the cube of the directions from ω to μ_i. Letting p and r coincide, the angle p, ω, q is then thrice the angle p, ω, μ_i, when

$$3p^2q+p^2+2pq-2(2p+q)-3=0$$

If in particular we take $p=\infty$, then $q=-1/3$; or if we take $p=0$, then $q=-3/2$. Fig. 47 is drawn for the latter case,

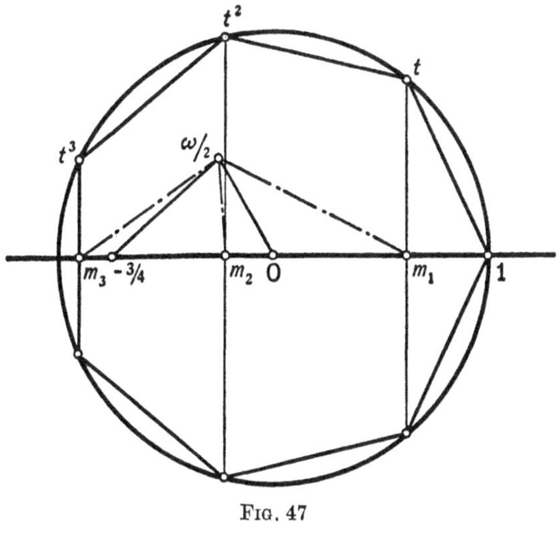

FIG. 47

the points $m_i = \mu_i/2$ being found by trisecting the angle $0, \omega/2, -3/4$.

Exercise 2 – The value of κ is here $(2\omega^2-1)/(2\omega-1)$.

Exercise 3 – Consider n a prime number of the form 2.3^r+1; for example, construct the 19-gon by two trisections.

Exercise 4 – Consider n a prime number of the form $2.2^r.3^s+1$; for example, $n=2.2.3+1=13$, for which the construction is effected by one bisection and one trisection. In this case there are six points μ_i where $\mu_1=\mu_{12}$, etc., $\mu_6=\mu_7$, etc., and $\sum^6 \mu_i+1=0$. Take the trisection first. Let

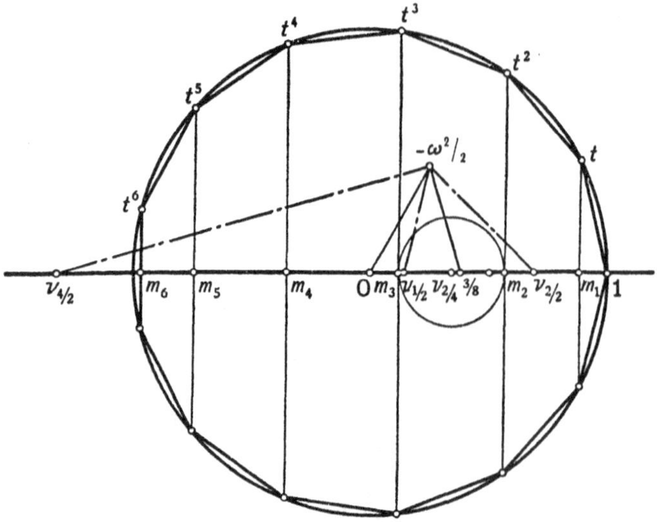

FIG. 48

$\nu_1=\mu_1+\mu_5$, $\nu_4=\mu_6+\mu_4$, $\nu_2=\mu_3+\mu_2$. To find $\nu_1/2$, $\nu_2/2$, $\nu_4/2$ trisect, say angle $0, -\omega^2/2, 3/8$. A pair of m's, say $m_2=\mu_2/2$, $m_3=\mu_3/2$, may then be identified by bisection, or equally (as in fig. 48) by cutting the base-line with a circle, centre $\nu_2/4$, radius $\sqrt{(\nu_2/4)^2-\nu_1/4}$.

§ 91. **The Regular 11-gon** – The equation to be solved is

(1) $$t^5+1/t^5+t^4+1/t^4+t^3+1/t^3+t^2+1/t^2+t+1/t+1=0$$

If t is any root, then t^2, t^4, t^8, $t^{16}\equiv t^5$, t^{10}, t^9, t^7, t^3, t^6 are also roots. Let

$$\mu_1=t+1/t=t^{10}+1/t^{10}$$
$$\mu_2=t^2+1/t^2=t^9+1/t^9$$
$$\mu_4=t^4+1/t^4=t^7+1/t^7$$
$$\mu_3=t^8+1/t^8=t^3+1/t^3$$
$$\mu_5=t^5+1/t^5=t^6+1/t^6$$

We have

$$\begin{aligned}
\mu_1^5 &= t^5 + 1/t^5 + 5(t^3 + 1/t^3) + 10(t + 1/t)\\
\mu_1^4 &= t^4 + 1/t^4 + 4(t^2 + 1/t^2) + 6\\
\mu_1^3 &= t^3 + 1/t^3 + 3(t + 1/t)\\
\mu_1^2 &= t^2 + 1/t^2 + 2\\
\mu_1 &= t + 1/t
\end{aligned}$$

whence μ_1 is a root of

(2) $$\mu^5 + \mu^4 - 4\mu^3 - 3\mu^2 + 3\mu + 1 = 0$$

and since we can replace t by t^2, t^4 . . . the roots of (2) are μ_i. We have $\mu_1^2 = \mu_2 + 2$, $\mu_2^2 = \mu_4 + 2$, . . ., $\mu_5^2 = \mu_1 + 2$, and generally

(3) $$\mu_r\mu_s = \mu_{r+s} + \mu_{r-s}, \quad \text{where} \quad \mu_0 = 2$$

Let now ϵ be a fifth root of unity, say $\epsilon = 1^{1/5}$. Consider the expressions:

$$\begin{aligned}
v_1 &= \mu_1 + \epsilon\mu_2 + \epsilon^2\mu_4 + \epsilon^3\mu_3 + \epsilon^4\mu_5\\
v_2 &= \mu_1 + \epsilon^2\mu_2 + \epsilon^4\mu_4 + \epsilon\mu_3 + \epsilon^3\mu_5\\
v_3 &= \mu_1 + \epsilon^3\mu_2 + \epsilon\mu_4 + \epsilon^4\mu_3 + \epsilon^2\mu_5\\
v_4 &= \mu_1 + \epsilon^4\mu_2 + \epsilon^3\mu_4 + \epsilon^2\mu_3 + \epsilon\mu_5
\end{aligned}$$

where the last three are obtained from the first by writing ϵ^2, ϵ^3, ϵ^4 for ϵ. Since the μ_i are real, $\bar{v}_1 = v_4$ and $\bar{v}_2 = v_3$. Squaring v_1 and reducing by (3) we have

$$\begin{aligned}
v_1^2 = \mu_2 + 2 &+ \epsilon^2(\mu_4 + 2) + \epsilon^4(\mu_3 + 2) + \epsilon(\mu_5 + 2) + \epsilon^3(\mu_1 + 2)\\
&+ 2\epsilon(\mu_3 + \mu_1) + 2\epsilon^2(\mu_5 + \mu_3) + 2\epsilon^3(\mu_4 + \mu_2) + 2\epsilon^4(\mu_5 + \mu_4)\\
&\qquad + 2\epsilon^3(\mu_5 + \mu_2)\\
&+ 2\epsilon^4(\mu_5 + \mu_1) + 2(\mu_4 + \mu_3) + 2(\mu_4 + \mu_1) + 2\epsilon(\mu_2 + \mu_1)\\
&\qquad + 2\epsilon^2(\mu_3 + \mu_2)
\end{aligned}$$

or collecting terms in μ_i and reducing each coefficient to three terms by

$$\epsilon^4 + \epsilon^3 + \epsilon^2 + \epsilon + 1 = 0$$

we have

$$v_1^2 = (2\epsilon - 2\epsilon^2 - \epsilon^3)(\mu_1 + \epsilon^2\mu_2 + \epsilon^4\mu_4 + \epsilon\mu_3 + \epsilon^3\mu_5)$$

or

(4) $$v_1^2 = (2\epsilon - 2\epsilon^2 - \epsilon^3)v_2$$

It follows without calculation that

(5) $$v_2^2 = (2\epsilon^2 - 2\epsilon^4 - \epsilon)v_4$$

and taking the conjugates

$$\bar{v}_2{}^2 = v_3{}^2 = (2\epsilon^3 - 2\epsilon - \epsilon^4)\bar{v}_1$$
$$\bar{v}_1{}^2 = v_4{}^2 = (2\epsilon^4 - 2\epsilon^3 - \epsilon^2)\bar{v}_2$$

The number $(2\epsilon - 2\epsilon^2 - \epsilon^3)$ has as modulus $\sqrt{11}$; it may be written as $\sqrt{11} \, . \, \tau_1$. Similarly $(2\epsilon^2 - 2\epsilon^4 - \epsilon)$ may be written as $\sqrt{11} \, . \, \tau_2$. Hence from (4) and (5), since $\bar{v}_1 = v_4$,

$$v_1{}^4 = (\sqrt{11})^3 \tau_1{}^2 \tau_2 \bar{v}_1$$

and similarly

$$v_2{}^4 = (\sqrt{11})^3 \tau_2{}^2/\tau_1 \, . \, \bar{v}_2$$

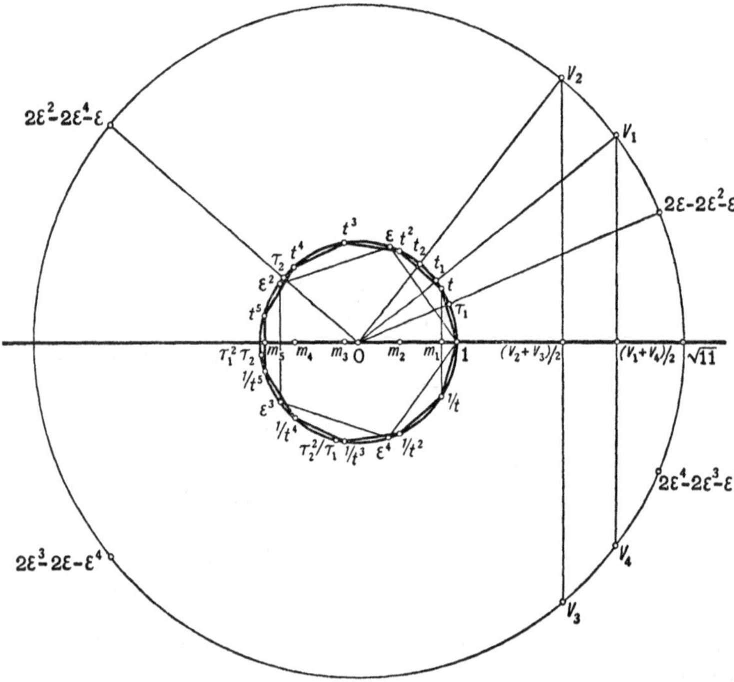

FIG. 49

If v_1 be written as ρt_1, v_2 as ρt_2, we have

$$\rho^3 t_1{}^5 = (\sqrt{11})^3 \tau_1{}^2 \tau_2$$
$$\rho^3 t_2{}^5 = (\sqrt{11})^3 \tau_2{}^2/\tau_1$$

Then $\rho = \sqrt{11}$, and t_1, t_2 are given by

$$t_1{}^5 = \tau_1{}^2 \tau_2$$
$$t_2{}^5 = \tau_2{}^2/\tau_1$$

These equations may be solved by dividing two given angles each into five equal parts. Having selected a value for t_1, the corresponding value for t_2 is determined without further ambiguity by the connection

$$t_2 = t_1^2/\tau_1$$

The point μ_1 is determined by v_1, v_2 and their conjugates v_4, v_3. For adding

$$-1 = \mu_1 + \mu_2 + \mu_3 + \mu_4 + \mu_5$$

to the four expressions v_i, we have as the sum

$$5\mu_1 = v_1 + v_2 + v_3 + v_4 - 1$$

To recapitulate the construction geometrically, it presupposes the construction of the regular pentagon 1, ϵ, ϵ^2, ϵ^3, ϵ^4. With this given the two points $(2\epsilon - 2\epsilon^2 - \epsilon^3)$, $(2\epsilon^2 - 2\epsilon^4 - \epsilon)$ may be marked on the circle centre 0, radius $\sqrt{11}$. Radii to these points cut the base-circle at τ_1, τ_2; one may thus mark the points $\tau_1^2\tau_2$, τ_2^2/τ_1 on the base-circle. Divide the positive arcs from 1 to $\tau_1^2\tau_2$, τ_2^2/τ_1 each into five equal parts. Choose the first partition in each case, as the determination of t_1, t_2. Produce radii through t_1, t_2 to meet the outer circle at v_1, v_2. Perpendiculars from v_1, v_2 to the base-line mark the real points $\frac{1}{2}(v_1 + v_4)$, $\frac{1}{2}(v_2 + v_3)$. The point $\mu_1/2$ or m_1 on the base-line is determined by

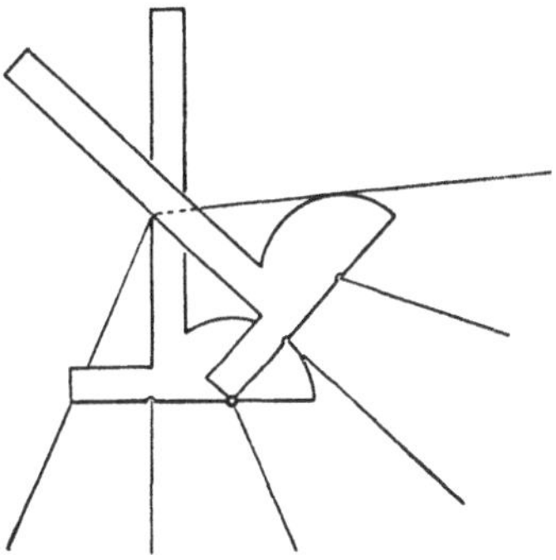

FIG. 50

$$5m_1 = \tfrac{1}{2}(v_1 + v_4) + \tfrac{1}{2}(v_2 + v_3) - \tfrac{1}{2}$$

The perpendicular to the base-line at m_1 cuts the base-circle at t, the first vertex of the regular 11-gon (fig. 49).

Exercise 5 – Two equal trisectors provide a simple instrument for dividing an angle into five parts, and so on (fig. 50).

§ 92. **Knots** – The construction of a regular pentagon by tying a knot in a strip of paper leads to a generalisation for the construction of any regular polygon of $2n+3$ sides.*

The construction of the pentagon is as follows (fig. 51 (*a*)). Call the ends of the strip α, β. Holding β horizontal,

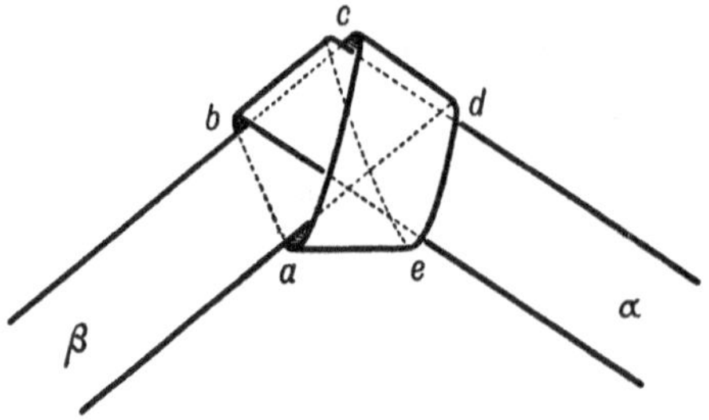

FIG. 51 (*a*)

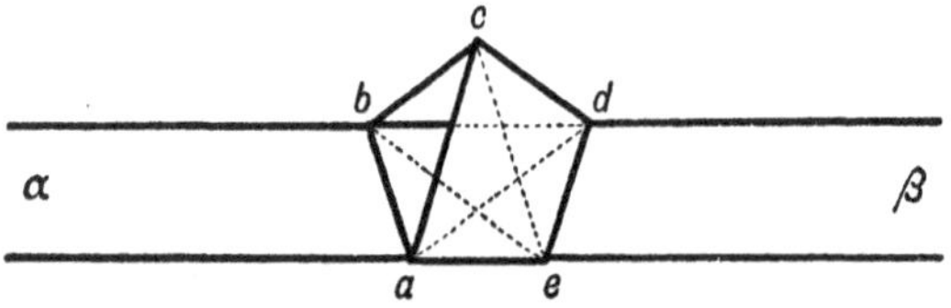

FIG. 51 (*b*)

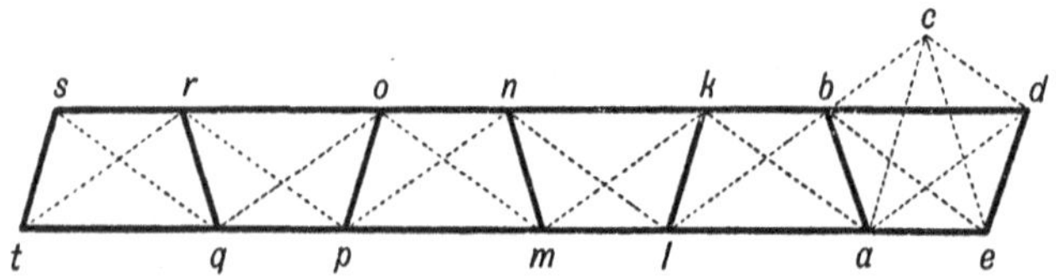

FIG. 51 (*c*)

(1) Fold α up on *cd*;

(2) Fold α down on *ea*, carrying it under *dabc* (that is, under β);

(3) Fold α up on *bc*, carrying it under *acde* (that is, through the loop).

Pull ends $\alpha_1\beta_1$ and when the knot is tight, flatten by creasing *cd*, *ea*, *bc*.

* For the pentagonal knot, *cf.* Lucas, *Récréations Mathématiques*, vol. 2, p. 202, and Faurey, *Procédés Originaux de Constructions Géométriques* (Paris, 1924), p. 135. For the generalisation, see F. V. Morley, *Proc. Lond. Math. Soc.*, June 1923.

Two further steps are possible:

(4) Fold *a* down on *de*, carrying it under *ceab*, to emerge at *ab*;

(5) Fold *β* up on *ab*, carrying it under *acde*, to emerge at *de*.
Flatten these last folds by creasing *de, ab* (fig. 51 (*b*)). The knot is completely tied. The ends *a*, *β* are collinear, but present different faces.

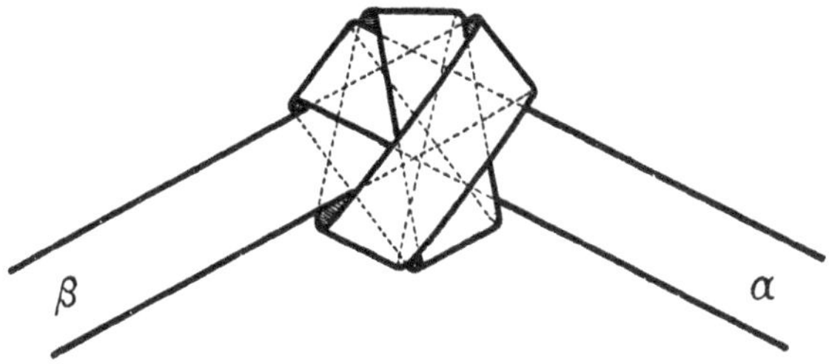

FIG. 52 (*a*)

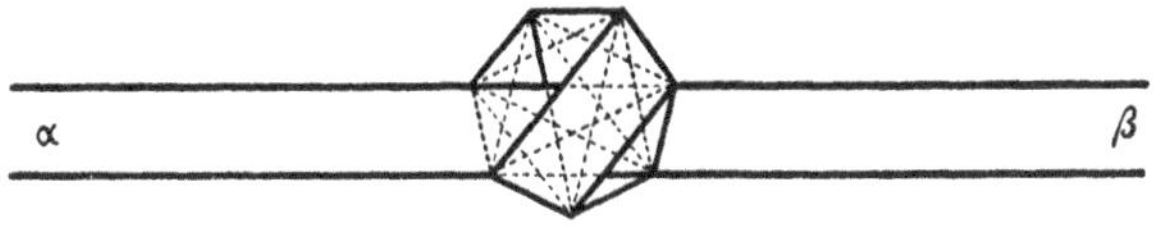

FIG. 52 (*b*)

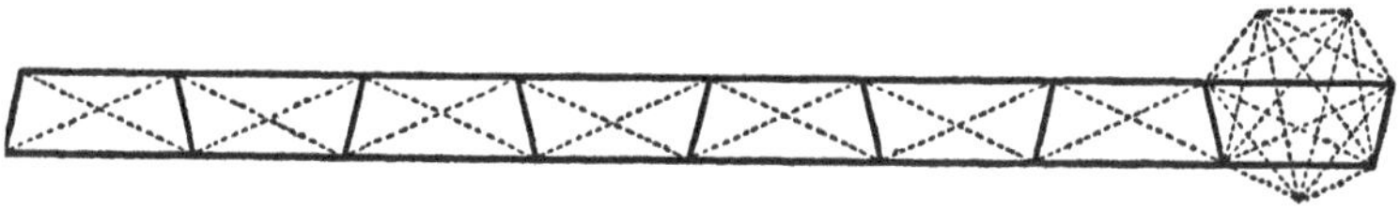

FIG. 52 (*c*)

If, with a somewhat narrower strip of the same length, we start from stage (3) in the above process, and, instead of pulling the ends immediately, pass *a* again down and below the double thickness, we have a double loop instead of a single one. By carrying *a* up and through the double loop, and then by pulling both ends and flattening when the knot is tight, we have the regular heptagon (fig. 52 (*a*)). Two further steps make the knot completely tied (fig. 52 (*b*)).

In this manner we may construct any odd regular polygons

of a greater number of sides merely by increasing the loopage of the knot.

Exercise 6 – The theorem will include the equilateral triangle if we admit the following process as knotting. Call the ends of the strip α, β and the edges 1, 2. Holding β horizontal, fold α up on bc, letting edge 2 cross edge 1 at a point a; fold α down on ca, so that edge 1 comes under b.

Exercise 7 – Construct the even regular polygons by knotting two strips.

CHAPTER XIV

MOTIONS

§ 93. **The Equation of a Motion** – The general rotation is given by

(1) $$y = tx - b$$

We regard this now not as an alibi, but as an alias. We think of a plane whose points are named y moving over a plane whose points are named x, the unit of length being the same in both planes. The relation (1) expresses that the point y is on the point x. It is a restatement of the equations for change of origin and rectangular axes in elementary analytic geometry.

From the ∞^3 equations (1) we select ∞^1 by writing $b = f(t)$. We have then for varying t the equation of a motion

(2) $$y = tx - f(t)$$

If we fix x, the equation maps the circle $|t| = 1$ on to the points of the y-plane which successively coincide with x; that is, the path traced on the y-plane by a pin fixed in the x-plane. Similarly, if we fix a pin in the y-plane we have in (2) the map-equation of the path traced in the x-plane.

If we make slots along two such paths we have two pins moving in two slots, and the motion is mechanically defined. The two pins may be in one plane, and therefore the two slots in the other, or there may be a pin and a slot in each plane.

Motions then are classified by the nature of the function $f(t)$. When the function is rational the motion is rational.

§ 94. **The Point of No Velocity** – For a given x, say x_0, we have a path in the y-plane, for which

$$dy = x_0 dt - f'(t) dt$$

Thus when $x_0 = f'(t)$, the point x_0 has no velocity over the y-plane. The path then has a cusp.

There is then in the x-plane a locus of points of no velocity,

(1) $$x_0 = f'(t)$$

This is called the centrode in the x-plane.

So for a given y, say y_0, we have a path in the x-plane for which

$$x = y_0/t + f(t)/t$$
$$dx/dt = -y_0/t^2 + f'(t)/t - f(t)/t^2$$

and the point y_0 has no velocity in the x-plane when

(2) $$y_0 = tf'(t) - f(t)$$

This curve is the centrode in the y-plane

Since

$$y_0 = tx_0 - f(t)$$

the points of no velocity are at any instant superposed.

For any constant x,

$$dy = (x - x_0)dt$$

and for any constant y,

$$dx = -(y - y_0)dt/t^2$$

For superposed x and y, $y - y_0 = t(x - x_0)$, so that

(3) $$dy/(y - y_0) = -dt/t = -dx/(x - x_0)$$

Since dt/t is a pure imaginary, $\iota d\theta$, dy is normal to $y - y_0$, and dx to $x - x_0$. Hence in the motion the paths of x and y touch.

The motion is thus the sliding of the x-paths $y = b$ over the y-paths $x = a$ (or conversely) where b and a obey the equation

$$b = ta - f(t)$$

The motion may be described as the rolling of the centrodes on each other. For, from (1) and (2)

$$dx_0 = f''(t)dt$$
$$dy_0 = tf''(t)dt$$

whence not only do the centroids touch, but also the elements of arc are equal.

But this is no simplification mechanically. For example, to make a circle roll on a circle is a fundamental mechanical problem which is solved by making toothed wheels which slide on one another.

§ 95. **The Points of No Acceleration** – To write the velocity and acceleration we need a real variable θ. Let then $t = e^{\iota\theta}$, or $dt/d\theta = \iota t$. The velocity of x over the y-plane is then

$$\dot{y} \equiv dy/d\theta = \iota t dy/dt = \iota t(x - x_0)$$

The acceleration is

$$\ddot{y} \equiv d^2y/d\theta^2 = \iota t d\dot{y}/dt = -t\{x - f'(t) - tf''(t)\}$$

There is then a point of no acceleration given by

(1) $$x_1 = f'(t) + tf''(t)$$

or by

$$x_1 - x_0 = tf''(t)$$

where x_0 is the point of no velocity.

Thus the acceleration of x is

(2) $$\ddot{y} = -t(x - x_1)$$

The velocity of a given y over the x-plane is, since $x = y/t + f(t)/t$,

$$\dot{x} = \iota t dx/dt = -\iota\{y/t - f'(t) + f(t)/t\} = -\iota(y - y_0)/t$$

The acceleration is

$$\ddot{x} = \iota t d\dot{x}/dt = -y/t - tf''(t) + f'(t) - f(t)/t$$

Thus there is a point of no acceleration given by

(3) $$y_1 = tf'(t) - f(t) - t^2 f''(t)$$

Exercise 1 – These points of no acceleration are then not superposed; they are opposite with regard to the point of no velocity.

§ 96. **The Curvature of a Path** – We here prove the curvature formula used. When x describes a curve, given in terms of a real parameter which we take as the time, then

$$\dot{x} = \dot{s}e^{\iota\phi}$$

where s is the arc, and ϕ the angle of the curve at the point considered.

The acceleration is the stroke

(1) $$\ddot{x} = \ddot{s}e^{\iota\phi} + \iota\dot{s}e^{\iota\phi}\dot{\phi}$$

The radius of curvature is the stroke

$$r = \iota \dot{s} e^{\iota\phi}/\dot{\phi}$$

Hence

$$\ddot{x}\bar{r} = -\iota \dot{s}\ddot{s}/\dot{\phi} + \dot{s}^2$$

and

$$\ddot{x}\bar{r} + \bar{\ddot{x}}r = 2\dot{s}^2$$

That is,

(2) $$|\ddot{x}|\ |r|\cos(\ddot{x}, r) = \dot{s}^2 = |\dot{x}|^2$$

We now find the centre of curvature of a path. Let (fig. 53) δ be the distance from x to x_0—that is, since $\dot{y} = \iota t(x - x_0)$, let δ be $|\dot{y}|$. Let κ be the distance from x_0 to x_1. Let ψ be the included angle x, x_0, x_1. The distance from x to x_1 is, by § 95 (2), $|\ddot{y}|$. Hence, by the curvature formula (2),

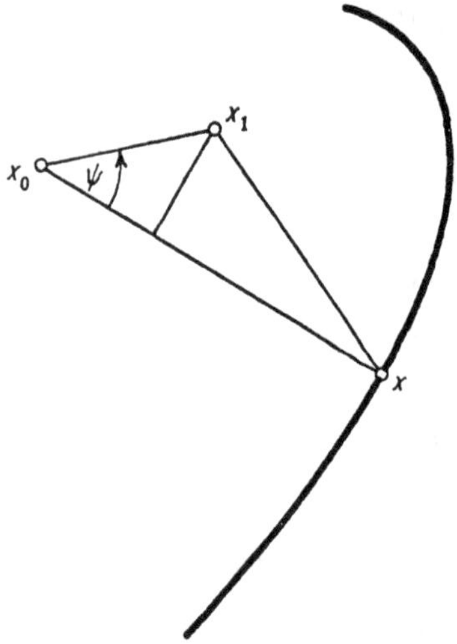

Fig. 53

(3) $$\delta^2/r = \delta - \kappa \cos\psi$$

or, if $r = \delta + \delta'$,

$$\delta\delta'/|r| = \kappa \cos\psi$$

or

(4) $$(1/\delta + 1/\delta')\cos\psi = 1/\kappa$$

The latter formula shows that if y' be the centre of curvature of the path of x, then y is the centre of curvature of the path of x'.

The path of x has a point of inflexion when $\delta' = \infty$, that is when

$$\delta = \kappa \cos\psi$$

This is the circle of which x_0 and x_1 are diametral. It is called the *circle of inflexions*.

There is a circle of inflexions in each plane, as shown in fig. 54.

Exercise 2 – The number $1/\kappa$ is the relative curvature of the two centrodes at the point of contact—that is, the difference of their curvatures.

Granting this (4) takes the form

$$(1/\delta + 1/\delta')\cos\varphi = 1/R + 1/R'$$

for the curvature of a point when a curve rolls externally on another curve. This is Savary's formula.

We have in (4) the curvature of the path $x = \text{constant}$. For the path $y = \text{constant}$ we must write $\psi - \pi$ for ψ. The formula is then

$$(1/\delta + 1/\delta'') \cos \psi = -1/\kappa$$

Hence

$$2/\delta + 1/\delta' + 1/\delta'' = 0$$

or measuring δ from x_0

(5) $$2/\delta = 1/\delta' + 1/\delta''$$

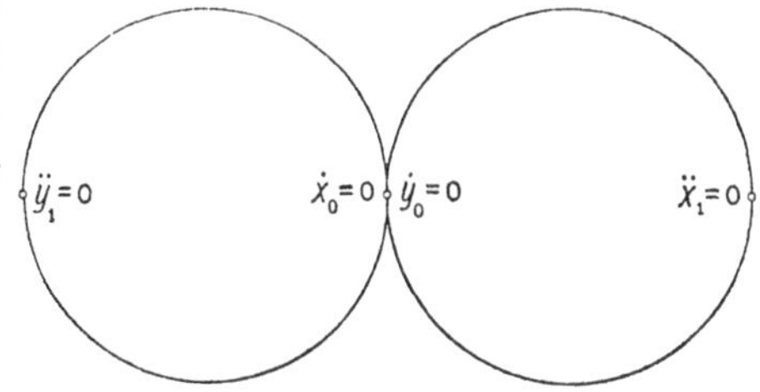

FIG. 54

That is, the two centres of curvature are apolar with x_0, and the overlying points x, y.

Thus when a circle in the x-plane rolls on a line in the y-plane the path of the centre x of the circle is a line. When then the line rolls on the circle the point y which overlies x describes a path whose centre of curvature is the mean of x and the point of contact.

Exercise 3 – The equation of this motion is

$$y = xt - \log t$$

By the theorem the path of $y = 0$ should have at $x = 0$ the centre of curvature $x = 1/2$.

Verify this analytically.

§ 97. **Envelopes** – In a given motion the points of a curve $x = \phi(\tau)$ in the x-plane will mark slots in the y-plane which will in general have an envelope. We have y a function of two turns t and τ,

$$y = t\phi(\tau) - f(t)$$

By the rule of envelopes (§ 84) we have that

$$t\{\phi(\tau) - f'(t)\}/\tau t \phi'(\tau) \text{ is real.}$$

When the curve $x = \phi(\tau)$ is a path; that is, when

$$\phi(\tau) = b/t + f(t)/t$$

the point b will be a part of the envelope.

§ 98. **The da Vinci Motion** – For a simple illustration of motions, let $f(t)$ be t^2; that is, consider the quadratic

(1) $$y = xt - t^2$$

The points of no velocity are

$$x_0 = 2t, \qquad y_0 = t^2$$

The loci of these—the centrodes—are

$$x = 2t, \qquad y = t^2$$

Thus the motion is the rolling of these two circles on each other.

For given y we have a slot in the x-plane,

(2) $$x = t + y/t$$

Thus $y = 1$ gives the straight slot $x = t + 1/t$, and $y = t_0$ gives the straight slot $x = t + t_0/t$. The motion is that of two pins in the y-plane moving in two straight slots in the x-plane. In this form it is the elliptic trammel, the general path in the x-plane being the ellipse.

The general path in the y-plane is a limaçon; but for $x = 0$ it is a circle.

Exercise 4 – Draw limaçons by cutting perpendicular straight slots in a card, and placing them over fixed pins. A pencil whose point passes through a small hole in the card describes the curve.

Suppose that we wish to draw the curve

$$y + a_0 t^3 + a_1 t^2 + a_2 t = 0$$

Replacing a_0 by x we get an equation of motion. For $x = 0$ we have the limaçon

$$y + a_1 t^2 + a_2 t = 0$$

And for $y = 0$ we have a related limaçon

$$x + a_1/t + a_2/t^2 = 0$$

Thus a pin in each plane engages a slot in the form of a limaçon in the other plane.

Step by step, then, we can draw the curve *

$$y + a_0 t^n + a_1 t^{n-1} + \quad . \quad . \quad . \quad + a_{n-1} t = 0$$

Exercise 5 – Let $y = xt - t/(1 - t)$

Show that the motion is that caused by a pin in each plane moving in a straight slot in the other plane, the distance from pin to slot in each plane being the same.

Exercise 6 – Let $y = tx - t^n$

Then the path in either plane is a trochoid, the centrodes being circles.

---

* F. Morley, *American Journal of Math.*, vol. 47, 1925.

Exercise 7 – If in a triangle ABC the points B, C are fixed and the angles B, C are in a given ratio, then if the edge BA is fixed any point attached to CA describes a trochoid.

§ 99. **Three-bar Motion** – Another simple example is provided by the jointed quadrangle, the lengths of whose sides in order of size are ρ_1, ρ_2, ρ_3, ρ_4, and their directions t_i. We have then two conjugate identities expressing closure:

$$(1) \qquad \rho_1 t_1 + \rho_2 t_2 + \rho_3 t_3 + \rho_4 t_4 = 0$$

$$(2) \qquad \rho_1/t_1 + \rho_2/t_2 + \rho_3/t_3 + \rho_4/t_4 = 0$$

When one bar ρ_4, t_4 of the jointed quadrangle is fixed there is still one degree of freedom, which gives three-bar motion. Let $\rho_4 = 1$, $t_4 = -1$ by calling the ends of the fixed bar 0 and 1. The identities are then

$$(3) \qquad \sum^3 \rho_i t_i = 1$$

$$(4) \qquad \sum^3 \rho_i / t_i = 1$$

The tracing-point x, rigidly attached to the traversing bar ρ_2, t_2, may be named from the base-point 0 as

$$(5) \qquad x_0 = \rho_1 t_1 + \kappa \rho_2 t_2$$

where κ is constant. Named from the point 1, x will be (using (3))

$$x_1 = \rho_2 t_2 (\kappa - 1) - \rho_3 t_3$$

Thus x is best named symmetrically from base-point anywhere as

$$(6) \qquad x = c_1 \rho_1 t_1 + c_2 \rho_2 t_2 + c_3 \rho_3 t_3$$

where the c_i are constants. Considering each c_i as a point of the plane, since (3) holds, we may write (6) as

$$x - c_3 = (c_1 - c_3)\rho_1 t_1 + (c_2 - c_3)\rho_2 t_2$$

which must be identical with (5) when $c_3 = 0$, $c_1 = 1$. Thus

$$c_1 - c_3 = 1$$
$$c_2 - c_3 = \kappa$$

There is then a third point κ which plays a part equivalent to 0 or 1. Hence we have the triple generation theorem of

Roberts, that the curve traced by a three-bar linkwork attached to two fixed points c_3, c_1 will also be traced simultaneously by three-bar linkworks attached to the other pairs of fixed points c_1, c_2 and c_2, c_3.

Exercise 8 – If

$$\sum^{3} \rho_i t_i = 1, \qquad \sum^{3} \rho_i / t_i = 1$$

$$\sum^{3} c_i \rho_i t_i = x, \qquad \sum^{3} c_i \rho_i / t_i = y$$

then

$$xy + \bar{x} + \bar{y} = \mathrm{A}$$

This equation and its conjugate give the focal pairing (see § 109) which sends the three-bar curve into itself. For further discussion of the three-bar curve see Hippisley, *Proc. Lond. Math. Soc.*, series 2, vol. 18, p. 136; Bennett, *ibid.*, vol. 20, p. 73; F. V. Morley, *ibid.*, vol. 21, p. 140.

The three-bar curve will be bipartite, unipartite, or rational (fig. 55) according as the coefficients $\rho_2 + \rho_3$ are greater than, less than, or equal to $\rho_1 + \rho_4$. If we force the two conditions

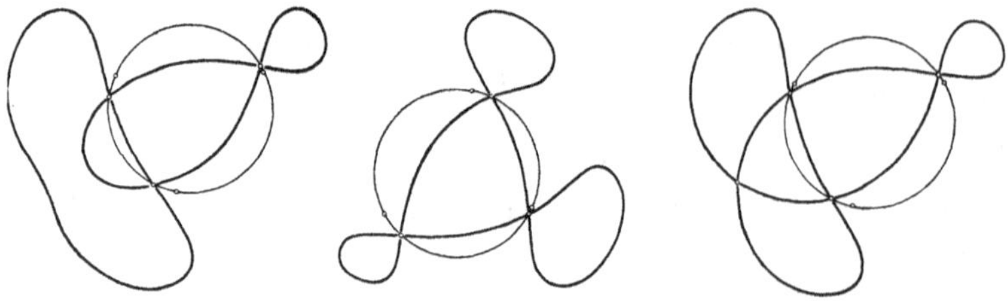

FIG. 55

to be true

$$-\rho_1 + \rho_2 + \rho_3 - \rho_4 = 0$$
$$-\rho_1 + \rho_2 - \rho_3 + \rho_4 = 0$$

which give $\rho_1 = \rho_2$ and $\rho_3 = \rho_4$, the jointed quadrangle may take either of the arrangements:

(1) Parallelogram or contraparallelogram, in which the equal bars are opposite.

(2) Kite, in which the equal bars are adjacent.

The triple generation figure shows that when one three-bar is in parallelogram form, the other two are kites. Hence the three-bar motion is in each case the same.

We analyse the second case. With ρ_4 fixed with ends 0 and 1 the identities are:

$$\rho_1(t_1+t_2)=1-t_3$$
$$\rho_1(t_1+t_2)=t_1t_2(1-1/t_3)$$

whence $t_3=1$, or $t_3=-t_1t_2$. In the former we have collapse of the bar ρ_3 on the fixed bar, and a point p attached to ρ_2 will trace a circle about 0. In the latter consider the plane of ρ_2, the x-plane, as sliding over the plane of the pivots, the y-plane. The naming of p in the y-plane is

$$y=\rho_1t_1+\rho_1\kappa t_2$$

and in the x-plane is

$$x=\rho_1\kappa$$

The equation of motion is

$$y=\rho_1t_1+t_2x$$

or, in terms of t,

$$y=\rho_1t+x(1-\rho_1t)/(\rho_1-t)$$

For given x this is a biquadratic with a node.

For the x-centrode $D_ty=0$, giving the limaçon

$$(\rho_1^2-1)x=\rho_1(\rho_1-t)^2$$

The y-centrode is another limaçon

$$y(\rho_1^2-1)=\rho_1^2-2\rho_1t+\rho_1^2t^2$$

It is geometrically obvious that in the case of contraparallelogram motion the centrodes are either ellipses or hyperbolas according to whether a short or a long side of the contraparallelogram is fixed. The triple generation property points out the equivalence between the rolling of two conics and the rolling of two limaçons.

The da Vinci motion is a special case of three-bar motion where ρ_1, ρ_3, ρ_4 are all infinite.

Exercise 9 – The kite or parallelogram motion, the da Vinci motion, and the connecting-rod motion (where ρ_1 and ρ_4 are infinite) are cases of

$$\rho_2^2+\rho_4^2=\rho_1^2+\rho_3^2$$

which is the condition that a rhombus may be inscribed with two opposite vertices on one circle, the other two on another circle.

Exercise 10 – When the products of opposite sides of the jointed quadrangle are equal, we have a poristic arrangement of an equilateral octagon on two circles.

CHAPTER XV

THE TRIANGLE

§ **100. The Nine-point Circle** – In a study of the triangle we take the circumcircle as base-circle. The vertices are then three turns, t_1, t_2, t_3, or collectively t_i.

We write

$$s_1 = t_1 + t_2 + t_3$$
$$s_2 = t_2 t_3 + t_3 t_1 + t_1 t_2$$
$$s_3 = t_1 t_2 t_3$$

These are the product-sums or elementary symmetric functions of t_i.

Their conjugates are:

$$\bar{s}_1 = s_2/s_3$$
$$\bar{s}_2 = s_1/s_3$$
$$\bar{s}_3 = 1/s_3$$

The mid-point of t_2 and t_3 is

$$x = (t_2 + t_3)/2$$
$$= (s_1 - t_1)/2$$

Hence

(1) $$x = (s_1 - t)/2$$

will pick up the three mid-points. And it is of the form $x = a + bt$; thus we have the map-equation of the circle on the mid-points. Its centre is $s_1/2$ and its radius is half that of the base-circle. It is called the nine-point or the Feuerbach circle of the triangle.

We call the process here illustrated of passing from points to a curve on them, *interpolation*.

§ **101. The Orthocentre** – The equation of a line on t_1 and perpendicular to the chord t_2, t_3 is

$$x - \bar{x} t_2 t_3 = t_1 - t_2 t_3 / t_1$$

or

$$x t_1 - \bar{x} s_3 = t_1 s_1 - s_2$$

This gives the altitude-line on t_i. In the interpolation

(1) $$xt - \bar{x}s_3 = ts_1 - s_2$$

All three altitude-lines are included.

Since (1) vanishes for any t when $x = s_1$, $\bar{x} = s_2/s_3$, the altitude-lines meet at the point

(2) $$x = s_1$$

This point is called the *orthocentre* of the triangle. Since each point t_1, t_2, t_3, s_1 is the orthocentre of the other three, the four form what is called an orthocentric set or four-point.

To find where an altitude-line meets the opposite edge, we write the equation of an edge—that is, the line joining t_2 and t_3. It is

$$x + \bar{x}t_2t_3 = t_2 + t_3$$

The image of t_1 in this line is

$$x = t_2 + t_3 - t_2t_3/t_1$$

The mid-point of this and t_1 is

$$x = (s_1 - t_2t_3/t_1)/2$$

This then is the foot of the altitude-line. Now, since t_2t_3/t_1 is a turn, this is a point on the nine-point circle.

Exercise 1 – The nine-point circle belongs to the orthocentric set, not merely to the triangle. It should then be on the mid-point of s_1 and t_1. Prove this.

Exercise 2 – Prove that the image of the orthocentre in an edge is on the circumcircle.

Exercise 3 – Prove that the four circumcircles of an orthocentric set are equal.

Exercise 4 – Construct, say with a meccano set, six jointed rhombuses of 0, t_1, t_2, t_3, t_1+t_2, t_2+t_3, t_3+t_1, $t_1+t_2+t_3$ (as in fig. 3). Verify that it is deformable in its plane. If the joints are universal it can become a cube.

§ **102. The Centroid** – The *centroid* or centre of gravity of the three given points is

$$x = s_1/3$$

It is obtained by taking the mid-point of t_2 and t_3,

$$m = (t_2 + t_3)/2$$

and dividing the segment formed by m and t_1 (in this order) in the ratio 1 : 2.

Since the circumcentre is 0, the centroid $s_1/3$, the nine-point centre $s_1/2$, and the orthocentre s_1, these points are on a line, called the *Euler line*, of the triangle. It is a convenient base-line in the discussion of a triangle. That is, we can when we wish take s_1 to be a positive number p.

Exercise 5 – If the clinant of the Euler line is c_4, and the clinants of the given lines c_i, then

$$c_4(c_1+c_2+c_3)+c_2c_3+c_3c_1+c_1c_2=0$$

The centroid divides the circumcentre 0 and the orthocentre s in the ratio 1 : 2. It is the centre of a stretch which sends the vertices on to the edges of the triangle.

This suggests that we find the general homology $y=a+bx$, which will send the triangle into another whose vertices are on the sides of the first. That is, to find a and b when $a+bt_1$ is on

$$x+\bar{x}t_2t_3=t_2+t_3$$

and similarly for t_2 and t_3.

We have then three equations,

$$a+bt_1+(\bar{a}+\bar{b}/t_1)t_2t_3=t_2+t_3=s_1-t_1$$

or

$$(a+bt_1)t_1{}^2+(\bar{a}t_1+\bar{b})s_3=s_1t_1{}^2-t_1{}^3$$

Interpolating, we have the cubic equation

$$(1+b)t^3+(a-s_1)t^2+\bar{a}s_3t+\bar{b}s_3=0$$

which is identical with

$$(1+b)(t^3-s_1t^2+s_2t-s_3)=0$$

Hence

$$(1+b)s_1=s_1-a$$
$$(1+b)s_2=\bar{a}s_3$$
$$(1+b)s_3=-\bar{b}s_3$$

Therefore

$$b+\bar{b}+1=0$$

When the homology is a stretch, we have $b=\bar{b}=-1/2$,

$$a=s_1/2, \qquad y=(s_1-x)/2$$

Thus the centre is $x=s_1/3$, the centroid.

When the homology is a rotation, then $b=t$, and

$$t+1/t+1=0$$

Thus the rotation is through $2\pi/3$.

Exercise 6 – The centre of the homology is on the circle of which the centroid and orthocentre are diametral points.

§ 103. **Euler's Relation** – We consider the triangle formed by three lines t_i of the base-circle (the tangents at the points t_i). The vertices are given by

$$\bar{x}_1 = 2/(t_2 + t_3)$$
$$= 2/(s_1 - t_1)$$

Hence the circumcircle of this triangle x_i is

(1) $$\bar{x} = 2/(s_1 - t)$$

It is then the inverse of the nine-point circle of t_i as to the base-circle.

Hence its centre and radius may be readily found (§ 86). But, keeping to Euclidean geometry, we seek the stretches which send the base-circle (an incircle of x_i) into the circumcircle. We have

$$\bar{x}_1 = 2(t_1 + t_2)(t_1 + t_3)/(t_2 + t_3)(t_3 + t_1)(t_1 + t_2)$$
$$= 2(s_2 + t_1^2)/(s_1 s_2 - s_3)$$

Take for simplicity s_1 as a positive number p. Then $s_2 = ps_3$, and

$$\bar{x}_1 = 2(p + t_1^2/s_3)/(p^2 - 1)$$

or

(2) $$x_1 = 2(p + s_3/t_1^2)/(p^2 - 1)$$

The point s_3/t_1^2 or $t_2 t_3/t_1$ is where the parallel to the secant t_2, t_3 meets the base-circle again. Calling it τ_1, the points τ_i are from (1) homothetic with x_i. The other homothetic set is $-\tau_i$, the points where the altitude-lines meet the base-circle again. The two stretches are

(3) $$x = 2(p \pm y)/(p^2 - 1)$$

The circumcentre of x_i is then $\delta = 2p/(p^2 - 1)$ and the radius is $R = \pm 2/(p^2 - 1)$.

We take as standard case (fig. 56) the base-circle as inscribed; the triangle t_i is thus acute-angled. In this case x_1, x_2, x_3 are in the same order as t_1, t_2, t_3, and $R = 2/(1 - p^2)$.

Thus

$$\delta^2 - R^2 = 4/(p^2 - 1) = -2R$$

or if the radius of the inscribed circle be a positive number r

(4) $$\delta^2 = R^2 - 2Rr$$

If the triangle be obtuse-angled, the base-circle is escribed and one of the radii, R or r, is negative.

Conversely when two circles satisfy Euler's relation (4), δ being the distance of the centres, then are ∞^1 triangles inscribed in the one and circumscribed to the other. The triangles are given by assigning s_1, the orthocentre of the

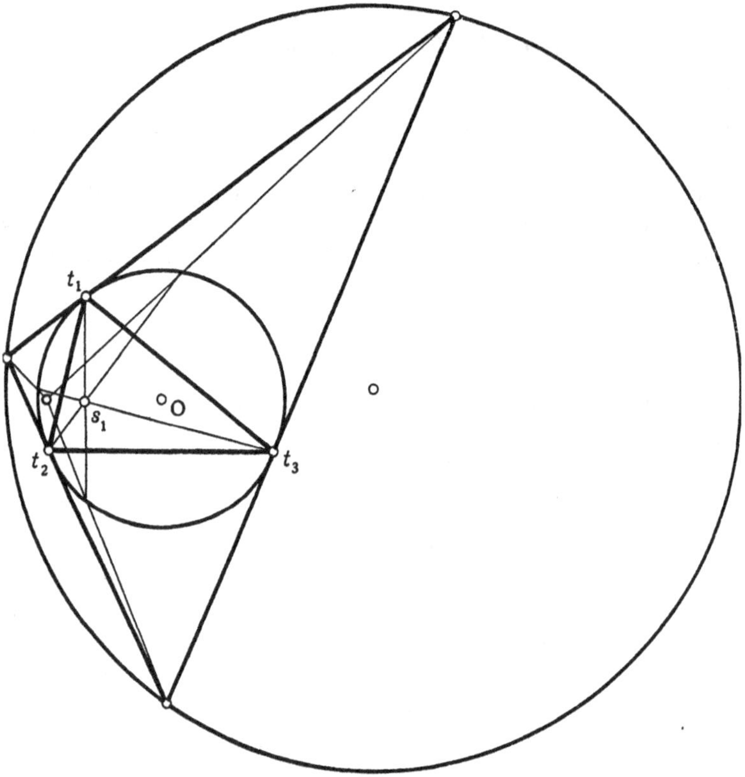

FIG. 56

points of contact. Thus with $s_1 = p$ they are given by the pencil of cubics

$$t^3 - pt^2 + s_3(pt - 1) = 0$$

Exercise 7 – The 2 : 2 correspondence of t_1 and t_2 is

$$t_1t_2(p^2 - 1) - p(t_1 + t_2)(1 + t_1t_2) + (t_1 + t_2)^2 = 0$$

Exercise 8 – An escribed circle B of a triangle has with the circumcircle C two common points and two common lines. Prove that the line of B on a common point, and the point of C on a common line, are incident.

Exercise 9 – The external bisectors of the angles of the triangle of tangents meet the opposite sides at points on the line $x\bar{s}_1 + \bar{x}s_1 = 6$ (J. H. Weaver, *Am. Math. Monthly*, vol. 40 (1933), p. 91).

Euler's relation is a relation between two circles and their radical axis. If the circles B, C intersect then the angles made by B, C, and the radical axis A are such that

$$\delta^2 = R^2 + 2Rr$$
$$\delta^2 = R^2 + r^2 - 2Rr \cos A$$

so that

$$2(1 + \cos A) = r/R = \sin B/\sin C$$

or

$$2 \sin C = \sin (A - C)$$

§ **104. Feuerbach's Theorem** – Let us determine the nine-point circle of the triangle of tangents t_i. The mid-point of x_2 and x_3 is, when $t_1 + t_2 + t_3 = p$,

$$m_1 = \tfrac{1}{2}(x_2 + x_3) = \{2p + s_3(1/t_2{}^2 + 1/t_3{}^2)\}/(p^2 - 1)$$

or, since

$$1/t_1{}^2 + 1/t_2{}^2 + 1/t_3{}^2 = p^2 - 2p/s_3$$
$$m_1 = s_3(p^2 - 1/t_1{}^2)/(p^2 - 1)$$

Writing $y = s_3/t^2$ we have one of the stretches which send the base-circle into the sought circle,

$$m = (s_3p^2 - y)/(p^2 - 1)$$

The centre is $m = s_3$, a point on the base-circle. Therefore the circles touch. That is, a circle which touches the lines of a triangle touches also the nine-point circle. This is Feuerbach's theorem (fig. 57).

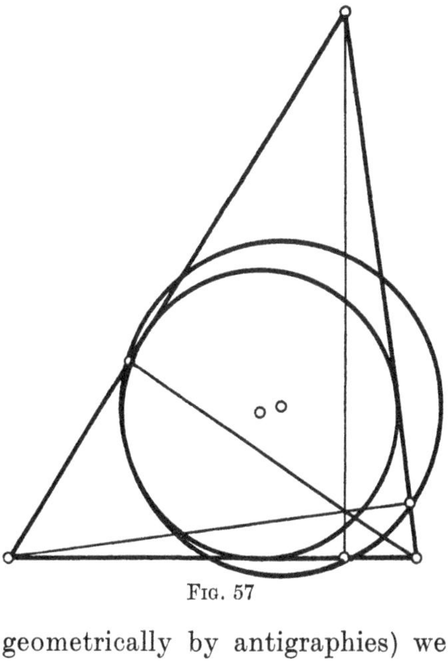

FIG. 57

§ **105. Interpolation** – From the three points t_i, by a finite number of the elementary operations (that is, algebraically by addition, subtraction, multiplication, division; geometrically by antigraphies) we can only obtain a rational fraction,

$$x = f(t_1, t_2, t_3)$$

(1) If x is symmetrical in all t's, then it can be expressed in terms of s_1, s_2, s_3. It is a covariant point for all permutations.

(2) If x is symmetrical in two t's, it is then one of a set of three points, obtained by a cyclic permutation. In this case we express x in terms of s_1, s_2, s_3 and the third t. Regarding this as a variable, we have a rational curve on the set. But using the cubic $t^3 - s_1t^2 + s_2t - s_3 = 0$ it can be replaced by either

(1) $$x = a + bt + ct^2$$

or

(2) $$x = a + bt + c/t$$

that is, by a limaçon or an ellipse on the set.

To obtain the circle on the set, let us by an homology write (1) as

$$x = bt + t^2$$

Then since

$$(t^2 + bt)(t - s_1 - b) + (b^2 + bs_1 + s_2)t - s_3 \equiv t^3 - s_1t^2 + s_2t - s_3$$

we have the circle on the set in the form

$$x(t - s_1 - b) + (b^2 + bs_1 + s_2)t - s_3 = 0$$

(3) If x has no symmetry, it is one of a set of six points, given by all permutations. To handle the six points we must use in addition to s_1, s_2, s_3 some alternating function such as

$$\mathrm{A} = (t_2 - t_3)(t_3 - t_1)(t_1 - t_2)$$

We have then, if

$$x_{123} = f(t_1, t_2, t_3)$$
$$x_{213} = f(t_2, t_1, t_3)$$

$x_{123} + x_{213}$ symmetric in t_1 and t_2, and therefore a function of t_3.

But

$$x_{123} - x_{213} = (t_1 - t_2) \times \text{a symmetric function in } t_1t_2$$
$$= \mathrm{A} \times \text{a symmetric function in } t_1t_2$$

so that x_{123} is by addition a function of t_3 alone, with coefficients in s_1, s_2, s_3, and A. It is reducible as before, but the coefficients will contain A in general. The six points fall into two sets of three; we may obtain the circle on each set.

§ **106. Taylor's Circle** – As an illustration of the third case, let the feet of the altitude lines be x_i and let the perpendicular from x_1 on t_3, t_1 meet it at x_{12}. Then

$$2x_1 = s_1 - t_2t_3/t_1$$

the image of this in the secant t_3, t_1 is

$$x = t_3 + t_1 - \bar{x}_1 t_3 t_1$$

so that

$$2x_{12} = x_1 + t_3 + t_1 - \bar{x}_1 t_3 t_1$$

and

$$4x_{12} = s_1 + at_1 - \Sigma t_2t_3/t_1$$

where a is the alternating function

$$a = 1 + t_1/t_2 + t_2/t_3 + t_3/t_1$$

Thus the points x_{12}, x_{23}, x_{31} are on the circle whose centre c is $\frac{1}{4}(s_1 - \Sigma t_2t_3/t_1)$ and radius is $\frac{1}{4}|a|$. The points x_{13}, x_{32}, x_{21} are given by

$$4x_{13} = s_1 + \bar{a}t_1 - \Sigma t_2t_3/t_1$$

and are therefore on the same circle. This is the Taylor circle of the triangle.

We have

$$(x_{12} - c)/a = (x_{13} - c)/\bar{a}$$

a rotation sending x_{12}, x_{23}, x_{31} into x_{13}, x_{21}, x_{32}.

Exercise 10 – The angle θ of this rotation is given by

$$\cot\theta = \cot a_1 \cot a_2 \cot a_3$$

Exercise 11 – For the Brocard angle we have

$$\cot\beta = \cot a_1 + \cot a_2 + \cot a_3$$

Thus these two angles, θ and β, specify the shape of a triangle.

§ **107. The Incentres** – In the problem of the circles which touch three lines, we have to bisect angles in order to find their centres. To avoid the root sign, let us replace t_i by τ_i^2. By τ_i is then meant a selected value of $\sqrt{t_i}$. Let us take the mid-points of the direct arcs as we go round the base-circle in the positive way. Let the unit-point be between τ_3^2 and τ_1^2, and let the mid-points of the direct arcs from 1 to τ_i^2 be τ_1, τ_2, τ_3. Then the mid-point of the arc τ_1^2 to τ_2^2 is $\tau_1\tau_2$, that of τ_2^2 to τ_3^2 is $\tau_2\tau_3$, but that of τ_3^2 to τ_1^2 is $-\tau_3\tau_1$. In this last case we have gone round the circle to reach τ_1^2, and the sign of τ_1 is changed. The in-centre is now the join of the chords

$$\begin{array}{ll} \tau_1^2, & \tau_2\tau_3 \\ \tau_2^2, & -\tau_3\tau_1 \\ \tau_3^2, & \tau_1\tau_2 \end{array}$$

so that

$$x = \tau_1\tau_2 + \tau_2\tau_3 - \tau_3\tau_1$$

Thus of the eight points

$$x = \pm\sqrt{t_2}\sqrt{t_3} \pm \sqrt{t_3}\sqrt{t_1} \pm \sqrt{t_1}\sqrt{t_2}$$

the incentres are those with an odd number of minus signs.

Exercise 12 – These four points are an orthocentric set. They are the counter-tetrad of ∞, t_i.

If we have four points on a circle, τ_i^2, we have four triangles. The centres of the inscribed circles are:

$$\begin{aligned} x_4 &= \tau_1\tau_2 + \tau_2\tau_3 - \tau_3\tau_1 \\ x_1 &= \tau_2\tau_3 + \tau_3\tau_4 - \tau_4\tau_2 \\ x_2 &= \tau_3\tau_4 - \tau_4\tau_1 + \tau_1\tau_3 \\ x_3 &= -\tau_4\tau_1 + \tau_1\tau_2 + \tau_2\tau_4 \end{aligned}$$

Hence

$$\begin{aligned} x_4 - x_1 &= (\tau_1 + \tau_4)(\tau_2 - \tau_3) \\ x_1 - x_2 &= (\tau_2 - \tau_1)(\tau_3 - \tau_4) \end{aligned}$$

The clinants, found by dividing each by its conjugate, are $-\tau_1\tau_2\tau_3\tau_4$ and $\tau_1\tau_2\tau_3\tau_4$. Thus the included angle is a right angle, and we have the theorem that the four centres of the inscribed circles form a rectangle (fig. 58).

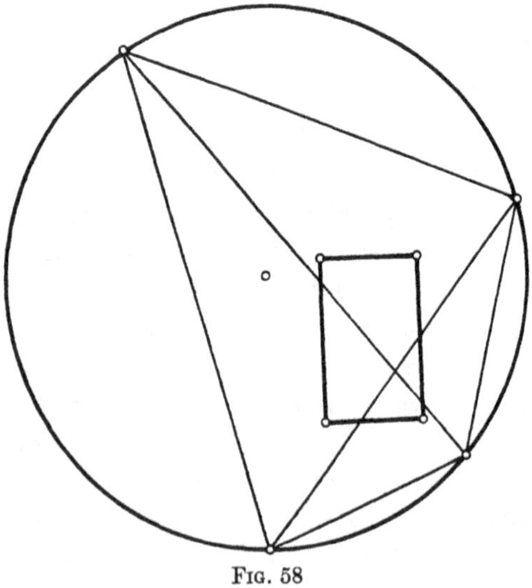

FIG. 58

Exercise 13 – Any edge of the rectangle and any pair of joins of the four points form an isosceles triangle.

Exercise 14 – The sixteen incentres are by sets of four on parallel lines, in two ways, one set of parallels being at right angles to the other.

§ **108. The Circle of Images** – The image of any point τ of the base-circle in a secant is

$$(1) \qquad \begin{aligned} x_1 &= t_2 + t_3 - t_2 t_3/\tau \\ &= s_1 - t_1 - s_3/\tau t_1 \end{aligned}$$

The three images b_i in the lines of a triangle are then included in

$$(2) \qquad x = s_1 - t - s_3/\tau t$$

This is for given τ the map-equation of a segment. Therefore the three images lie on a line, the *line of images*. The theorem variously ascribed to Wallace (*Mathematical Repository*, March 1799) and to Simson, that the feet of the perpendiculars are on a line follows at once.

The centre of the segment is s_1, the orthocentre. Thus all lines of images are on the orthocentre. In particular when $\tau = t_i$, the line is an altitude-line of the triangle.

Exercise 15 – For any multiples μ_i we have from (1)

$$\Sigma \mu_i b_i = s_1 \Sigma \mu_i - \Sigma \mu_i t_i - s_3 \Sigma \mu_i / t_i / \tau.$$

Hence if a_i be the angles of the triangle

$$\Sigma b_i \sin 2a_i = s_1 \Sigma \sin 2a_i.$$

Exercise 16 – If we apply to (2) the rule of envelopes, we obtain the orthocentre and the circle $x = s_1 - 2t$. This is the locus of cusps of the segments.

Let x_i be the images of any point x in the sides of the triangle t_i. We call the triangle x_i the triangle of images of x, and its circumcircle the circle of images of x.

Since

$$\begin{aligned} x_1 &= t_2 + t_3 - \bar{x} t_2 t_3 \\ 1 - \bar{x} x_1 &= (1 - \bar{x} t_2)(1 - \bar{x} t_3) \\ &= \Pi/(1 - \bar{x} t_1) \end{aligned}$$

where

$$(1) \qquad \Pi = 1 - s_1 \bar{x} + s_2 \bar{x}^2 - s_3 \bar{x}^3$$

Hence the circle of images is

$$(2) \qquad 1 - \bar{x} z = \Pi/(1 - \bar{x} t)$$

Since, when $z = \infty$, $t = 1/\bar{x}$, the centre y of this circle is given by

$$(3) \qquad 1 - \bar{x} y = \Pi/(1 - x\bar{x})$$

and the equation (2) is, subtracting (2) from (3)

$$z - y = \Pi(x - t)/(1 - x\bar{x})(1 - \bar{x}t)$$
$$= (1 - \bar{x}y)\tau$$

that is, the radius ρ is $| 1 - \bar{x}y |$, and

(4) $$\rho^2 = (1 - \bar{x}y)(1 - x\bar{y})$$

From (3)

$$(1 - x\bar{x})(1 - \bar{x}y) = 1 - s_1\bar{x} + s_2\bar{x}^2 - s_3\bar{x}^3$$

or

$$x + y - \bar{x}xy = s_1 - s_2\bar{x} + s_3\bar{x}^2$$

or

$$x + y - s_1 + s_3\bar{x}\bar{y} = \bar{x}[s_3(\bar{x} + \bar{y}) - s_2 + xy]$$

whence, if x is not a turn, we have

(5) $$x + y + s_3\bar{x}\bar{y} = s_1$$

and its conjugate

(5′) $$s_3(\bar{x} + \bar{y}) + xy = s_2$$

§ 109. **Focal Pairing** – We consider especially

(1) $$x + y + s_3\bar{x}\bar{y} = s_1$$

and regard it as a pairing, or transformation of period 2, of the points of the plane. It is one-to-one in general; but there are exceptions. For when $x = t_1$, then

$$y + t_2t_3\bar{y} = t_2 + t_3$$

Thus to a vertex t_1 corresponds any point on the opposite side.

And, when x passes to ∞ in the direction τ, then

$$\tau + s_3\bar{y}/\tau = 0$$

or y is a point on the base-circle, and for varying τ it is the base-circle.

The transformation thus sends each of the four points t_1, t_2, t_3, ∞ into the circle on the other three.

It has four fixed points, the incentres, given by

(2) $$2x + s_3\bar{x}^2 = s_1$$

Thus (1) is the polar form of (2).

The relations of x, y to t_i are as follows:—

(1) The angle-relations. The angle x, t_1, t_2 is the angle

t_3, t_1, y (mod π). That is, the clinant of $(x-t_1)(y-t_1)$ is the clinant of $(t_2-t_1)(t_3-t_1)$, or is $t_1{}^2t_2t_3$ or s_3t_1.

For suppose two points x, y such that

$$(x-t_i)(y-t_i)=s_3t_i(\bar{x}-1/t_i)(\bar{y}-1/t_i)$$

Then

$$t_i{}^3-t_i{}^2(x+y)+t_ixy-s_3[t_i{}^2\bar{x}\bar{y}-t_i(\bar{x}+\bar{y})+1]=0$$

or

$$t_i{}^2(x+y+s_3\bar{x}\bar{y}-s_1)-t_i[s_3(\bar{x}+\bar{y})+xy-s_2]=0$$

for three values of t_i. Whence we have again equation (1).

(2) The distance-relations. If ξ_i be the real distances from x to the three-lines, and η_i those for y, then, applying Ptolemy's theorem to the four points x, y, x_i, y_i we have

$$4\xi_i\eta_i=\rho^2-\delta^2$$

where δ is $|x-y|$. Or, from equation (4) of the last section,

$$4\xi_i\eta_i=(1-x\bar{y})(1-\bar{x}y)-(x-y)(\bar{x}-\bar{y})$$

$$=(1-x\bar{x})(1-y\bar{y}) \tag{2}$$

The product of distances of x and y from a line of the triangle is then a constant, and the constant is given by (2).

Assuming the elementary properties of conics, it is clear that x and y are foci of an inscribed conic.

The points x and y are usually called, from the angle-property, isogonal conjugates. The transformation of period 2, which interchanges the points, is then called transformation by isogonal conjugates. It is called by Pascal (*Repertorium, Geometry*, p. 732) the Desargues transformation.

In view of its inversive importance we shall call it focal pairing and speak of x and y as a focal pair.

The axes of the inscribed conic with foci x, y are given by

$$\rho^2=(1-x\bar{y})(1-\bar{x}y)$$

$$\rho^2-\delta^2=(1-x\bar{x})(1-y\bar{y})$$

The conic is a hyperbola, parabola, or ellipse as

$$\rho<\delta, \qquad \rho=\delta, \qquad \rho>\delta$$

Exercise 17 – For four points q_i in a Euclidean space, if the points x, y are foci of an inscribed quadric of revolution, then the angle which the plane q_1q_2x makes with $q_1q_2q_3$ is the angle which the plane $q_1q_2q_4$ makes with q_1q_2y.

Exercise 18 – Apply the transformation to the figure formed by the triangle and its medians.

§ **110. The Pedal Circles** – Let x and y be a focal pair. The pedal circle of x is the circle on the feet of the perpendiculars from x on the edges of the given triangle. Since its centre is the mid-point of x and y and its radius is half that of the circle of images, it is equally the pedal circle of y.

It has the equation

$$2z = x + y + (1 - y\bar{x})t$$

or

(1) $$2z = s_1 - s_3\bar{x}\bar{y} + (1 - y\bar{x})t$$

If $x=0$, $y=s_3$ and we have the special pedal circle

(2) $$2z = s_1 + t$$

which is the nine-point circle.

For any pedal circle let

$$-s_3\bar{x}\bar{y} - y\bar{x}t = 0$$

that is,

$$t = -s_3\bar{y}/y$$

Then $2z = s_1 + t$, a point on the nine-point circle. Thus the common points of (1) and (2) are discrete, being given by

$$2a = s_1 - s_3\bar{y}/y$$
$$2b = s_1 - s_3\bar{x}/x$$

Hence

$$(2a - s_1)/(2b - s_1) = x\bar{y}/y\bar{x}$$

or, in words, if x, y subtend at 0 an angle θ, then a, b subtend at $s_1/2$ the angle 2θ. In particular, if $\theta = 0$ or $\theta = \pi$, the points a, b coincide and the pedal circle touches the nine-point circle. This particular case is M'Cay's extension of Feuerbach's theorem (see Casey, *Sequel to Euclid*).

§ **111. The Invariant I_2** – Let us ask for points x whose triangle of images is perspective with t_i. The join of t_1 and $t_2 + t_3 - \bar{x}t_2t_3$ is

$$\begin{vmatrix} z & t_2 + t_3 - \bar{x}t_2t_3 & t_1 \\ \bar{z} & 1/t_2 + 1/t_3 - x/t_2t_3 & 1/t_1 \\ 1 & 1 & 1 \end{vmatrix} = 0$$

that is, with $s_1 = p$,

$$\begin{aligned} & z(p - tx/s_3 - 2/t) \\ & -\bar{z}(p - \bar{x}s_3/t - 2t) \\ & \qquad + 1/t - t = \bar{x}s_3/t^2 - xt^2/s_3 \\ & \qquad\qquad = \bar{x}(t - p + s_2/t) - x(1/t - p + \bar{s}_2 t) \end{aligned}$$

Hence, the coefficients of t, $1/t$, 1 being zero when $p \neq 0$,

$$z - x = \bar{z} - \bar{x}$$

that is, the join of z and x is parallel to the Euler line, and also

$$2z - \bar{z}\bar{x}s_3 - p = x - \bar{x}ps_3$$
$$2\bar{z} - zx/s_3 - p = \bar{x} - xp/s_3$$

Eliminating z and $\bar{z}$, we have either

(1) $$(x - \bar{x})x\bar{x} + \bar{x}^2 s_3 - x^2/s_3 + 2(\bar{x} - x) + p(\bar{x}s_3 - x/s_3) = 0$$

or $p = 0$, which says that for an equilateral triangle x is any point.

This equation may be obtained more easily by asking that the join of a focal pair x, y be parallel to the Euler line.

For then

$$x + y + \bar{x}\bar{y}s_3 = p$$
$$\bar{x} + \bar{y} + xy/s_3 = p$$

and

$$y = x - \rho$$

so that

(2) $$2x + \bar{x}^2 s_3 - p = \rho(1 + \bar{x}s_3)$$
$$2\bar{x} + x^2/s_3 - p = \rho(1 + x/s_3)$$

and

$$(2x + \bar{x}^2 s_3 - p)(1 + x/s_3) = (2\bar{x} + x^2/s_3 - p)(1 + \bar{x}s_3)$$

which is a form of (1).

The map-equation (2) is the simplest. It assigns to any real, ρ, an orthocentric four-point on the curve. Thus $\rho = 0$ gives the four incentres.

The curve, as defined by Neuberg, is

(3) $$\begin{vmatrix} \lambda_{23}\lambda_{14} & \lambda_{23} + \lambda_{14} & 1 \\ \lambda_{31}\lambda_{24} & \lambda_{31} + \lambda_{24} & 1 \\ \lambda_{12}\lambda_{34} & \lambda_{12} + \lambda_{34} & 1 \end{vmatrix}$$

with the usual meaning $\lambda_{ij} = (x_i - x_j)(\bar{x}_i - \bar{x}_j)$. To identify the two forms, note that (3) is also a biquadratic. It is on ∞ since, when λ_{i4} becomes ∞, the second column is proportional to the third. It is on the base-point, for which $\lambda_{14} = \lambda_{24} = \lambda_{34}$. It is on the points t_i, for when $x = t_1$, $\lambda_{14} = 0$

and $\lambda_{24}=\lambda_{21}$. It is on the incentres, for then, in the usual notation of trigonometry, $\lambda_{23}=a\ .\ .\ .,$

$$\begin{aligned}\lambda_{14}&=r^2+(s-a)^2\\&=\frac{s-a}{s}\{(s-b)(s-c)+s(s-a)\}\\&=bc(s-a)/s\end{aligned}$$

whence (3) vanishes.

These nine points are manifestly on the curve (2) as defined by focal pairs. The two are then identical.

The points x, t_1, t_2, t_3 are then mutually related. But further, these four and ∞ are mutually related. For under any inversion (3) written for cyclic interchange of 1, 2, 3 as

$$\sum^{3}(\lambda_{23}\lambda_{31}+\lambda_{34}\lambda_{12})(\lambda_{14}-\lambda_{24})$$

becomes

$$\mathrm{I}_2\equiv\sum^{3}(\lambda_{23}\lambda_{31}\lambda_{45}+\lambda_{34}\lambda_{35}\lambda_{12})(\lambda_{14}\lambda_{52}-\lambda_{15}\lambda_{42})=0$$

It is thus a sum of twelve terms of the type

$$\lambda_{12}\lambda_{23}\lambda_{34}\lambda_{45}\lambda_{51}$$

with the sign + or − according as we interchange an even number of suffixes or not.

We have then in I_2 an invariant involving five points a_i symmetrically. When four of the points are given, the locus of the fifth is a biquadratic, C^2.*

Exercise 19 – Since $\mathrm{I}_2=0$ for t_i, ∞, p, the curve for four points a_i is on the counter-tetrad.

Deduce that the Neuberg curve is on the incentres.

Exercise 20 – Find when the Euler lines of the four triangles formed by four points meet at a point.†

---

* For this invariant I_2, see F. Morley, *American Mathematical Monthly*, vol. 32 (1925). For the determination of the four points when C^2 is given, see Morley and Patterson, *American Journal*, vol. 52 (1930). For the five points a_i so related we have five curves C_i^2 by regarding in turn each a_i as movable. The five curves meet at a point a_6; and the relation of the six points a_i is a mutual one. See a paper by Mrs Dean, *American Journal*, vol. 52 (1930).

† R. C. Yates. See a note in *American Journal*, vol. 54 (1932).

CHAPTER XVI

INVARIANTS UNDER HOMOLOGIES

§ 112. **Constants under Translations** – Consider the effect of a translation T, $x = y + b$, on an n-point, given by

(1) $$f(x) \equiv a_0 x^n + n a_1 x^{n-1} + \ . \ . \ . + a_n$$

It becomes, let us say,

(2) $$a_0 y^n + n \beta_1 y^{n-1} + \ . \ . \ . + \beta_n$$

the coefficient a_0 being unaltered. We take it as 1. We seek functions of the coefficients a_i which are equal to the same functions of β_i. These will be the constants or absolute invariants of (1) under T. The n-point has a centroid, $-a_1$. We make this the base-point by writing $x = z - a_1$, that is, by removing the second term. Let this be

(3) $$z^n + \binom{n}{2} \kappa_2 z^{n-2} + \ . \ . \ . + \kappa_n$$

This will be the same whether derived from (1) or (2). Thus the κ_i are the sought constants.

By Taylor's theorem

$$f(x) = f(z - a_1) = f(-a_1) + z f'(-a_1) + \frac{z^2}{2\,!} f''(-a_1) + \ . \ . \ . + z^n$$

Thus

$$\kappa_2 = x^2 + 2a_1 x + a_2, \qquad \text{when } x = -a_1$$
$$= a_2 - a_1^2$$
$$\kappa_3 = x^3 + 3a_1 x^2 + 3a_2 x + a_3, \qquad \text{when } x = -a_1$$
$$= a_3 - 3a_1 a_2 + 2a_1^3$$

and generally

$$\kappa_r = a_r - r a_1 a_{r-1} + \binom{r}{2} a_1^2 a_{r-2} \ . \ . \ . + (-)^r (1 - r) a_1^r$$

Exercise 1 – A triangle, $x^3 + 3a_1 x^2 + 3a_2 x + a_3$, is equilateral or regular if $\kappa_2 \equiv a_2 - a_1^2 = 0$. If the roots be x_1, x_2, x_3 the condition is either

$$x_1 + \omega x_2 + \omega^2 x_3 = 0$$

or

$$x_1 + \omega^2 x_2 + \omega x_3 = 0$$

Exercise 2 – One point of the triangle is the centroid of the other two if

$$\kappa_3 \equiv a_3 - 3a_1a_2 + 2a_1^3 = 0$$

Exercise 3 – A four-point is a parallelogram if $\kappa_3 = 0$.

Under homologies a_0 is no longer constant. Replacing a_r by a_r/a_0 and clearing of fractions, we have the invariants

$$\kappa_2 = a_0a_2 - a_1^2 \quad \text{or} \quad \mu^2 \times \text{same},$$
$$\kappa_3 = a_0^2a_3 - 3a_0a_1a_2 + 2a_1^3 \quad \text{or} \quad \mu^3 \times \text{same},$$

from which the constants are determined by elimination of μ.

Thus in the case of the cubic or three-point there is the one constant

$$\gamma_0 = \kappa_2^3/\kappa_3^2$$

This is all we can have, for triangles have only ∞^2 shapes, and the number γ_0 can take ∞^2 values.

When two of the points coincide, we may take $a_2 = 0$ and $a_3 = 0$, so that $\kappa_2 = -a_1^2$, $\kappa_3 = 2a_1^3$, $\gamma_0 = -1/4$.

Thus the form has a repeated factor when

$$\kappa_6 \equiv 4\kappa_2^3 + \kappa_3^2 = 0$$

A convenient choice of the absolute invariant is now κ_6/κ_3^2; that is, we replace γ_0 by $1 + 4\gamma_0$, say by γ.

In terms of the roots x, we have

$$9\kappa_2/a_0^2 = \Sigma(x_2x_3 - x_1^2)$$
$$27\kappa_3/a_0^3 = \Pi(x_2 + x_3 - 2x_1)$$
$$27\kappa_6/a_0^6 = -\Pi(x_2 - x_3)^2$$

It is enough to verify the numerical factors. When $a_1 = 0$, that is, when $x_1 + x_2 + x_3 = 0$, the second is true. And when the x_i are 0, 1, -1 the first and third are readily verified.

Under the reflexion $x = \bar{y}$ the cubic

$$a_0x^3 + 3a_1x^2 + 3a_2x + a_3$$

becomes on taking the conjugate

$$\bar{a}_0y^3 + 3\bar{a}_1y^2 + 3\bar{a}_2y + \bar{x}_3$$

so that all our invariants acquire values which are the conjugates of their old values. In particular γ becomes $\bar{\gamma}$. Thus γ is not an absolute invariant for the Euclidean group unless it happens to be real. If

$$\gamma = \rho e^{\iota\theta} = \xi + \iota\eta$$

ρ and ξ are absolute, whereas θ and η can change in sign.

When γ is real, the triangle admits a reflexion into itself; that is, it is either isosceles or flat (the three points being in a line).

To distinguish these cases, we observe that when the triangle is flat we can take the x_i as all real. Then κ_3^2/a_0^6 is positive, and κ_6/a_0^6 is negative. Thus γ is negative for a flat triangle.

But for an isosceles triangle we may take

$$x_1 = \rho, \qquad x_2 = \iota\rho', \qquad x_3 = -\iota\rho_3'$$

so that κ_3^2/a_0^6 is positive and κ_6/a_0^6 is positive. Thus γ is positive for an isosceles triangle.

Four points under homographies have two invariants, g_2 and g_3 (§ 30). The quartic can be taken with one point at ∞ as

$$4x^3 - g_2x - g_3$$

When the points are concyclic, this regarded as a cubic has its three zeros on a line.

Since here

$$a_0 = 4, \qquad a_2 = -g_2/3, \qquad a_3 = -g_3$$
$$\kappa_2 = -4g_2/3, \qquad \kappa_3 = -16g_3, \qquad \kappa_6 = -4^4(g_2^3/27 - g_3^2)$$
$$\gamma = -\frac{4^4}{3^3}(g_2^3 - 27g_3^2)/g_3^2$$

Therefore when the points are concyclic

$$(g_2^3 - 27g_3^2)/g_3^2 \text{ is positive.}$$

And when the points are anticyclic, the zeros of the cubic are isosceles, and

$$(g_2^3 - 27g_3^2)/g_3^2 \text{ is negative.}$$

§ **113. Lagrange Resolvents** – A polynomial in two variables x_1 and x_2 is a constant under translations T when the variables appear only in the combination $x_1 - x_2$. But conversely when it is a constant under T the variables must appear in this way. For if we write

$$v = x_1 - x_2, \qquad s = x_1 + x_2$$

so that

$$2x_1 = s + v, \qquad 2x_2 = s - v$$

then the polynomial is a polynomial in s and v, and if it is a constant it cannot contain s. The complete system is then simply v.

For polynomials in three variables x_1, x_2, x_3 we write

$$v_1 = x_1 + \omega x_2 + \omega^2 x_3$$
$$v_2 = x_1 + \omega^2 x_2 + \omega x_3$$

These are constants under T, since

$$1 + \omega + \omega^2 = 0$$

Let also

$$s = x_1 + x_2 + x_3$$

Then, solving for x_i, a polynomial in the x_i is a polynomial in s, v_1, v_2. If it is a constant it cannot contain s. Thus v_1 and v_2 are a complete system for polynomials which are constant under T.

In the same way, for n ordered points the $n-1$ expressions

$$v_i = x_1 + \epsilon x_2 + \epsilon^2 x_3 + \ . \ . \ . \ + \epsilon^{n-1} x_n$$

where ϵ is a root of $\epsilon^n = 1$ (other than $\epsilon = 1$), are a complete system for polynomials in x_i, under T. We call these $n-1$ expressions the Lagrange resolvents of the n-point.

Under homologies H these numbers v_i acquire a common factor. They are a complete system of relative invariants. There are $n-2$ constants, which may be taken as

$$v_2/v_1, \ . \ . \ . \ v_{n-1}/v_1.$$

Let us consider, in terms of these resolvents, some special ordered n-points.

(i) If an n-point is a point repeated, then

$$v_1 = v_2 = \ . \ . \ . \ = v_{n-1} = 0$$

Conversely if all Lagrange resolvents vanish, the ordered n-point is a repeated point.

(ii) If an n-point is regular, then we may take $x_1 = 1$, $x_2 = \epsilon$, . . . $x_n = \epsilon^{n-1}$. All resolvents vanish except v_{n-1}. And conversely.

(iii) For a next case take the points as given by

$$x_1 + a + b = 0, \qquad x_2 + a\epsilon + b\epsilon^2, \qquad x_3 + a\epsilon^2 + b\epsilon^4 \ . \ . \ .$$

For simplicity consider five points. Then $v_1 = 0$, $v_2 = 0$, $v_3 = -5b$, $v_4 = -5a$. All resolvents but the last two vanish.

Exercise 4 – Collectively the points are $x^5+a^5+b^5-5ab^2x^2+5a^3bx=0$.

Such a set of points is the images of a point x_0 in the edges of a regular polygon. They lie on the curve $x+at+bt^2=0$, which is a limaçon. If we reflect the point 1 in the lines $x+\bar{x}t^2=2\mu t$, the tangents of the circle $(0, \mu)$, we have the limaçon in the form

$$x=2\mu t-t^2$$

The node on double point is the point to which correspond two values of t, say t and t'. That is, $2\mu=t+t'$, $tt'=1$, and hence $x=1$. The focus is the point for which two values of t (where t is no longer restricted) are equal. That is, it is μ^2. The limaçon referred to its node is

$$x+1-2\mu t+t^2=0$$

or in polar co-ordinates

$$r=2\mu-2\cos\theta$$

Thus the n-point in this binomial case is constructed also by taking a point on a circle, joining it to the vertices of a regular polygon in that circle, and producing each join by the distance 2μ.

(iv) For a fourth case, let the points x_i lie on successive edges of a regular polygon. Then we may write

$$x_1=t_1(1+\mu_1\iota), \qquad x_2=\epsilon t(1+\mu_2\iota)\ \ldots$$

where the μ_i are real. Taking 5 points,

$$v_1=t\iota(\mu_1+\mu_2\epsilon^2+\mu_3\epsilon^4+\ \ldots)$$
$$v_2=t\iota(\mu_1+\mu_2\epsilon^3+\mu_3\epsilon\ +\ \ldots)$$

Hence

$$-t^2=v_1/\bar{v}_2=v_2/\bar{v}_1$$

Exercise 5 – The equation $v_1\bar{v}_1=v_2\bar{v}_2$ is, for any x_i, the equation of a line. This line is the edge of a regular pentagon on the other four points.

And so in general for an n-gon,

$$v_1/\bar{v}_{n-3}=v_2/\bar{v}_{n-4}=\ \ldots\ =v_{n-3}/\bar{v}_1$$

are the conditions on points taken on its successive edges, and the common value is the clinant of the first edge.

If this clinant is not definite then

$$v_1=v_2=\ \ldots\ =v_{n-3}=0$$

There are then on the points two regular n-gons.

Thus another definition of the third case is that the points are the intersection of corresponding sides of co-ordered regular n-gons—that is, where the first side meets the first side, and so on. There is a pencil of regular n-gons on the points, including as a special case the node of the limaçon. The circles inscribed in the n-gons are in fact bitangents to the limaçon, whence it follows that these circles are all orthogonal to a fixed circle.

§ **114. The Case of the Triangle** – Suppose the cubic for the three points x_i is

(1) $$x^3 - 3abx + a^3 + b^3$$

or

$$\begin{vmatrix} x & a & b \\ b & x & a \\ a & b & x \end{vmatrix}$$

with the factors

$$x_1 + a + b$$
$$x_2 + \omega^2 a + \omega b$$
$$x_3 + \omega a + \omega^2 b$$

Here

$$v_1 = x_1 + \omega x_2 + \omega^2 x_3 = -3a, \qquad v_2 = -3b$$

To construct v_i, we construct the equilateral triangle $x_1' x_2 x_3$. If it be outwards, then

$$x_1' + \omega x_3 + \omega^2 x_2 = 0$$

so that

$$v_2 = x_1 - x_1'$$

If we do the same for the other edges, we have three vectors

$$v_2, \qquad \omega v_2, \qquad \omega^2 v_2 \quad \text{(fig. 59)}$$

Let us now find the point or points at which the edges of the triangle make the same angle (mod π). One case is when

$$\frac{x - x_1}{\bar{x} - \bar{x}_1} = \omega \frac{x - x_2}{\bar{x} - \bar{x}_2} = \omega^2 \frac{x - x_3}{\bar{x} - \bar{x}_3}$$

whence each

$$= \frac{3x}{-\bar{x}_1 - \omega^2 \bar{x}_2 - \omega \bar{x}_3} = \frac{x_1 + \omega^2 x_2 + \omega x_3}{\bar{x}_1 + \omega \bar{x}_2 + \omega^2 \bar{x}_3}$$

that is,

(2) $$x = b\bar{a}/\bar{b} = f_2 \text{ say.}$$

The other case is found by writing ω^2 for ω, and gives

(2′) $$x = a\bar{b}/\bar{a} = f_1 \text{ say.}$$

These points occur in the solution of Fermat's problem, to find that point for which the sum of the distances from x

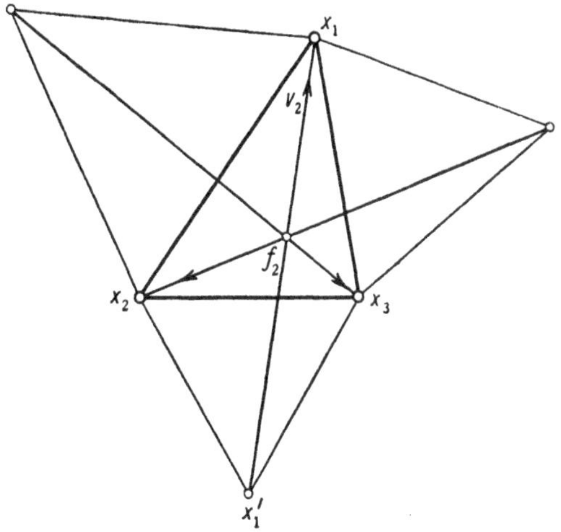

FIG. 59

to x_i is a minimum. For, if $\Sigma\sqrt{(x-x_i)(\bar{x}-\bar{x}_i)}$ is a minimum, then

$$\Sigma\sqrt{\frac{x-x_i}{\bar{x}-\bar{x}_i}} = 0, \qquad \Sigma\sqrt{\frac{\bar{x}-\bar{x}_i}{x-x_i}} = 0$$

But if

$$\alpha+\beta+\gamma=0 \quad \text{and} \quad 1/\alpha+1/\beta+1/\gamma=0$$

then

$$\alpha : \beta : \gamma = 1 : \omega : \omega^2 \quad \text{or} \quad = 1 : \omega^2 : \omega$$

We call the two points then the Fermat points of the triangle.

To prove that the three vectors of fig. 59 meet at f_2, we may either use Euclid's methods, or remark that

$$\begin{aligned} x_1 - f_2 &= -a - b - b\bar{a}/\bar{b} \\ &= -b(1 + b/a + \bar{b}/\bar{a}) \end{aligned}$$

a real multiple of b.

So the three vectors for the interior equilateral triangles meet at f_1.

From the equations (2) we see that the distance $|g-f_2|$ between the centroid g and the Fermat point f_2 is $|a|$; that is, $\frac{1}{3}|v_1|$, and $|g-f_1|$ is $\frac{1}{3}|v_2|$.

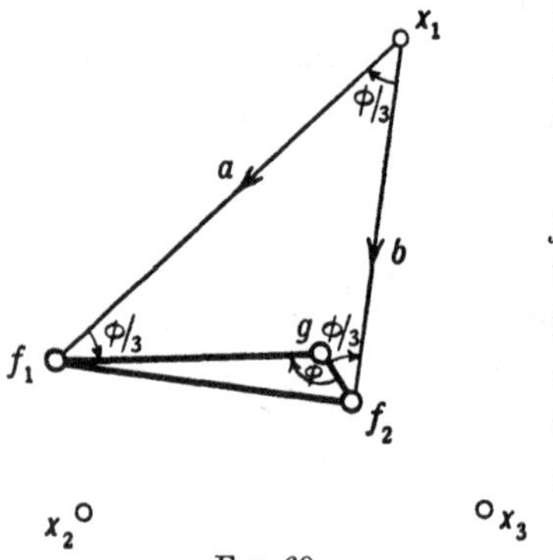

FIG. 60

But also the angle from b to f_2 is the angle from a to b, and the angle from a to f_1 is the angle from b to a. Hence if ϕ be the angle f_2, g, f_1, the angle from b to a (or from v_2 to v_1) is a third of ϕ (fig. 60).

A triangle is then uniquely constructed when the centroid and the Fermat points are given.

Exercise 6 – If p_1, p_2, p_3 are the squares of the edges and σ is the area taken as positive, then

$$v_1\bar{v}_1 = \tfrac{1}{2}(p_1+p_2+p_3) - 2\sqrt{3}\sigma$$
$$v_2\bar{v}_2 = \tfrac{1}{2}(p_1+p_2+p_3) + 2\sqrt{3}\sigma$$
$$v_1\bar{v}_2 = -p_1 - \omega^2 p_2 - \omega p_3$$

Exercise 7 – The angle ϕ is given by

$$e^{2\iota\phi} = (v_1\bar{v}_2/v_2\bar{v}_1)^3$$

The cubic

$$\begin{vmatrix} x & a & b \\ b & x & a \\ a & b & x \end{vmatrix}$$

is, as the sum of cubes,

$$[(ax-b^2)^3 \quad (bx-a^2)^3]/(a^3-b^3)$$

so that (§ 42) the Hessian points are

$$x = a^2/b = h_2, \qquad x = b^2/a = h_1$$

Let us write the antigraphies which send the points into themselves. For two of the points x and y we have

$$(ax-b^2)^3 = (bx-a^2)^3$$
$$(\bar{a}\bar{y}-\bar{b}^2)^3 = (\bar{b}\bar{y}-\bar{a}^2)^3$$

whence

(3) $$(ax-b^2)(\bar{a}\bar{y}-\bar{b}^2) = \epsilon(bx-a^2)(\bar{b}\bar{y}-\bar{a}^2)$$

where $\epsilon^3 = 1$. These three antigraphies interchange the Hessian points; for $\epsilon = 1$ we have the inversion in the circumcircle, and for $\epsilon = \omega$ or ω^2 we have the antigraphies $\overline{123}$ or $\overline{321}$ of § 39.

But also

(4) $$(ax - b^2)(\bar{b}\bar{y} - \bar{a}^2) = \Sigma(bx - a^2)(\bar{a}\bar{y} - \bar{b}^2)$$

For these each Hessian point is fixed, so that we have the three Apollonian inversions.

When in (3) we set $y = \infty$ we have

$$\bar{a}(ax - b^2) = \epsilon\bar{b}(bx - a^2)$$

Thus the circumcentre is

(5) $$x = (b^2\bar{a} - a^2\bar{b})/(a\bar{a} - b\bar{b})$$

and a Brocard point (§ 40) is

(6) $$x = \frac{b^2\bar{a} - \omega a^2\bar{b}}{a\bar{a} - \omega b\bar{b}}$$

Exercise 8 – The Hessian points subtend at a Brocard point the angle $2\pi/3$.

Let us now connect the Hessian points

$$h_1 = b^2/a, \qquad h_2 = a^2/b$$

with the Fermat points

$$f_1 = a\bar{b}/\bar{a}, \qquad f_2 = b\bar{a}/\bar{b}$$

We have

$$h_1\bar{f}_2 = b\bar{b}$$
$$h_2\bar{f}_1 = a\bar{a}$$

so that

g, h_1, f_2 are on a line

and

g, h_2, f_1 are on a line.

Also

$$h_1 - f_1 = (b^2\bar{a} - a^2\bar{b})/a\bar{a}$$

and

$$f_2 - h_2 = (b^2\bar{a} - a^2\bar{b})/b\bar{b}$$

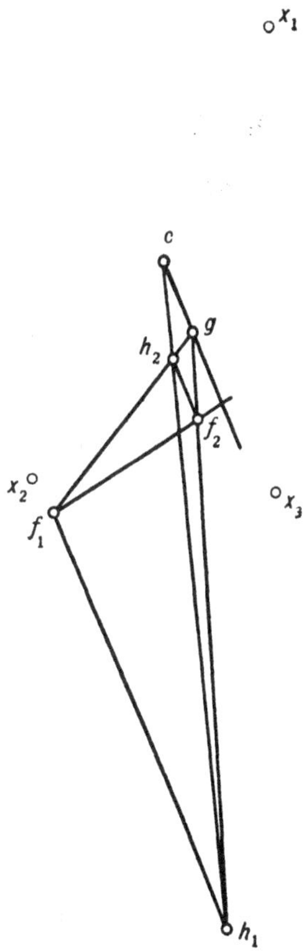

FIG. 61

These are then positively parallel. From (5) they are parallel to the join of the centroid and the circumcentre (fig. 61).

Exercise 9 – The intersection of the line on the Hessian points and that on the Fermat points is

$$\frac{b^2\bar{a}+a^2\bar{b}}{a\bar{a}+b\bar{b}}$$

This is the symmedian point.

§ **115. The Four-point** * – The Lagrange resolvents for an ordered four-point give a simple construction for the two circumscribed squares (fig. 62).

For by § 113 (iii) when $x_1'+\iota x_2-x_3-\iota x_4$ is 0 all squares described positively on x_2, x_3, x_4 will have the fourth edge on x_1'. If then we construct x_1'—that is, draw from x_3 a vector equal to x_4-x_2 and making with it the direction ι (or

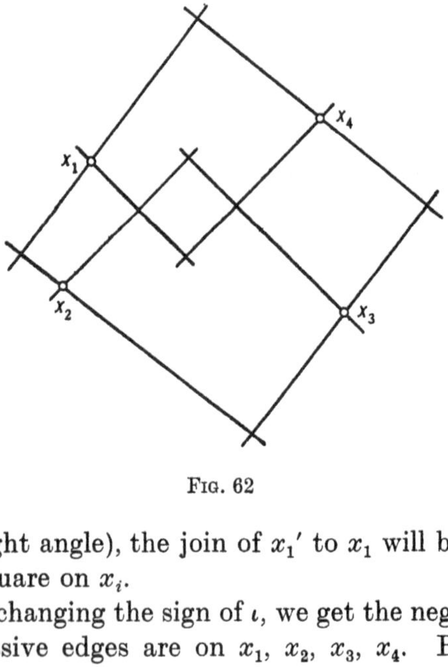

Fig. 62

a positive right angle), the join of x_1' to x_1 will be an edge of a positive square on x_i.

Similarly, changing the sign of ι, we get the negative square whose successive edges are on x_1, x_2, x_3, x_4. Both squares are shown in fig. 62.

So in the case of regular (or equilateral) triangles on three points, we have the positive series, containing the point f_2

* On the question of the complete system of comitants, under the Euclidean group, for a three-point and a four-point, see C. H. Rawlins, *American Journal*, vol. 40 (1918).

taken three times, and the negative series containing f_1 (§ 114). In the notation of § 114, the two limaçons (one for each series) are

$$x + at + bt^2 = 0$$

and

$$x + b\tau + a\tau^2 = 0$$

Taking the former, the conjugate is

$$\bar{b} + \bar{a}t + \bar{x}t^2 = 0$$

Eliminating t, we have the self-conjugate equation

$$(x\bar{a} - a\bar{b})(a\bar{x} - b\bar{a}) = (x\bar{x} - b\bar{b})^2$$

Writing similarly the second limaçon, we see by subtraction that the two meet at four points on a circle—that is, since they are on the points x_1, x_2, x_3 they meet again on the circumcircle.

Exercise 10 - If a_i, b_i, c_i, d_i is a positively ordered square when i is 1, 2, 3 or 4, and if a_1, b_1, c_1, d_1 and a_2, b_2, c_2, d_2 are also positively ordered squares, then a_3, b_3, c_3, d_3 and a_4, b_4, c_4, d_4 are positively ordered squares.*

For four points x_i under $y = tx + b$ we have

$$y_3 - y_1 = t(x_3 - x_1)$$
$$\bar{y}_2 - \bar{y}_4 = (\bar{x}_2 - \bar{x}_4)/t$$

so that

$$(x_3 - x_1)(\bar{x}_2 - \bar{x}_4) = \kappa$$

a constant. It is of course the product of the lengths of the diagonals and their relative direction, say $e^{\iota\phi}$.

Placing the join of the diagonals at the base-point, let

$$x_i = \rho_i e^{\iota\theta_i}$$

Then

$$\begin{aligned} -x_1\bar{x}_2 &= -\rho_1\rho_2 e^{\iota(\theta_1 - \theta_2)} \\ &= -\tfrac{1}{2}(\rho_1{}^2 + \rho_2{}^2 - p_{12}) + 2\iota\sigma_{012} \end{aligned}$$

where p_{12} is the squared distance of x_1, x_2 and σ_{012} is the area of the triangle 0, x_1, x_2.

And

$$\bar{x}_2 x_3 = \tfrac{1}{2}(\rho_2{}^2 + \rho_3{}^2 - p_{23}) + 2i\sigma_{023}$$

so that on the whole

$$\kappa = \xi + \iota\eta$$

---

* J. R. Musselman, "On Certain Types of Polygons," *American Math. Monthly*, vol. 40 (1933).

where

$$\xi = \tfrac{1}{2}(p_{12} - p_{23} + p_{34} - p_{41})$$
$$\eta = 2\sigma_{1234}$$

where σ_{1234} is the area of the quadrangle.

If for a polygon of an even number of sides we call

$$\tfrac{1}{4}(p_{12} - p_{23} + p_{34} - \ . \ . \ .)$$

the norm then for the quadrangle

$$\kappa = 2(\text{norm} + \iota \text{ area})$$

Exercise 11 – When the norm vanishes the diagonals are at right angles; when the area vanishes they are parallel.

Exercise 12 – The constant κ is not altered by translation of a diagonal relative to the other.

By interchange of suffixes we have for a four-point six invariants such as κ. Let us write them κ_i and $\bar{\kappa}_i$ where,

$$\kappa_1 = (x_2 - x_3)(\bar{x}_1 - \bar{x}_4)$$
$$\kappa_2 = (x_3 - x_1)(\bar{x}_2 - \bar{x}_4)$$
$$\kappa_3 = (x_1 - x_2)(\bar{x}_3 - \bar{x}_4)$$

The sum is 0. They may be taken as the complete system of invariants under homologies.

Exercise 13 – Neuberg's invariant (§ 111) is in this notation

$$\Sigma \kappa_i \bar{\kappa}_i(\kappa_i + \bar{\kappa}_i).$$

Exercise 14 – A canonical form for four points under homologies is $x_i = \mu_i + \iota/\mu_i$, where μ_i is real. Show that Neuberg's invariant vanishes when $3 + s_1 s_3 - 10 s_4 + 3 s_4^2 = 0$ where the s_i are product-sums of the μ_i.

§ 116. The Hexagon – When we have an equation $f = 0$ expressing a relation it is proper to ask for the meaning of f when it does not vanish.

Let us do this for the expression

(1) $$| \, x_i \quad \bar{y}_i \quad 1 \, |$$

whose vanishing means that the triangles x_i, y_i are negatively similar.

We regard the six points as vertices of a hexagon and write (1) as

$$\begin{vmatrix} x_1 & x_3 & x_5 \\ \bar{x}_4 & \bar{x}_6 & \bar{x}_2 \\ 1 & 1 & 1 \end{vmatrix}$$

that is

$$x_1\bar{x}_6 - \bar{x}_2x_1 + \ . \ . \ . \ - x_5\bar{x}_6$$

Now

$$x_1\bar{x}_6 = \tfrac{1}{2}(p_{01} + p_{06} - p_{16}) + 2\iota\sigma_{061}$$

where, as usual, p_{ij} is a squared distance and σ_{061} is the area of 0, x_6, x_1.

Thus the determinant is $\xi + \iota\eta$, where

$$2\xi = p_{12} - p_{23} + \ . \ . \ . \ - p_{61}$$

and

$$\eta = 2\sigma$$

where σ is the area of the hexagon.

The determinant is a constant under translations applied to either triangle. Under the homology $x = ax' + b$ applied to the triangle x_i it acquires the factor a. Thus by proper choice of a we can make either ξ or η zero.

If

$$\cot\theta = \xi/\eta = (p_{12} - p_{23} + \ . \ . \ . \ - p_{61})/4\sigma$$

then θ is a constant angle under translations of one triangle.*

Exercise 15 – If the hexagon reduces to a triangle by coincidence of x_6 and x_1, x_2 and x_3, x_4 and x_5, then θ becomes the Brocard angle given by

$$\cot\beta = (p_{23} + p_{31} + p_{12})/4\sigma$$

Exercise 16 – Given two ordered triangles x_1, x_3, x_5 and x_4, x_6, x_2, lines from the vertices of the second making with corresponding edges of the first the angle θ meet at a point.

Let us show in this case of the ordered hexagon how the Lagrange resolvents behave when n is not a prime. There are five resolvents, the first being

$$v_1 = x_1\omega^2x_2 + \omega x_3 - x_4 + \omega^2x_5 - \omega x_6$$

Let u_1 and u_2 be the resolvents for x_1, x_3, x_5 and u_1' and u_2 those for x_4, x_6, x_2

Then

$$v_1 = u_1 - u_1'$$
$$v_2 = u_2 + u_2'$$
$$v_3 = x_1 + x_3 + x_5 - (x_4 + x_6 + x_2)$$
$$v_4 = u_1 + u_1'$$
$$v_5 = u_2 - u_2'$$

* For the extension of this invariant to two polygons, see L. M. Blumenthal, *American Journal*, vol. 49 (1927).

An ordered n-gon, when the $n-1$ resolvents vanish, is a repeated point. For if $x_1 = x_2 = \ldots = x_n$ all resolvents vanish. When all but the last vanish, the n-gon is regular and in the positive sense. For if $x_1 = 1$, $x_2 = \epsilon$, $x_3 = \epsilon^3$, . . . all the v_i vanish except the last.

But if all but the last two vanish, we have a polygon which may be called semi-regular. The nature varies as n is odd or even.

In the present case of $n = 6$, we have $v_1 = v_2 = v_3 = 0$, that is,

$$u_1 = u_1'$$
$$u_2 = -u_2'$$
$$s_1 = s_1'$$

Taking the common centroid as base-point, we have by addition $\sqrt{3}x_4 = \iota(x_3 - x_5)$, etc.

From equation (2) of § 114, the interior Fermat points f_2 are the same. That is, the diagonals of the hexagon meet at a point, at angles $2\pi/3$. Also $u_2 + u_2' = 0$ means that the middle points m_i of the diagonals form a negative regular triangle.

Exercise 17 – The circumcircle of this triangle is on the point f_2. Its centre is the centroid. The point $-m_1$ is the centre of the positive regular triangle with vertices x_3 and x_5 (or x_6 and x_2).

Exercise 18 – The external Hessian points are the same for the two triangles.

Exercise 19 – A median of either triangle is $\sqrt{3}/2 \times$ the corresponding edge of the other.

Exercise 20 – The homography which sends x_1, x_3, x_5 into x_4, x_6, x_1 is

$$2s_2x_1x_4 = 3(s_3x_4 - s_3'x_1)$$

where

$$s_2 = \Sigma x_1x_3 = -s_2', \qquad s_3 = x_1x_3x_5, \qquad s_3' = x_4x_6x_2$$

This homography sends the circumcentre and symmedian point of the one triangle into the symmedian point and the circumcentre of the other.

§ 117. Barycentric Co-ordinates – We have for four points x_i the identity

$$|\, x - x_i, \quad x_i, \quad \bar{x}_i, \quad 1 \,| = 0$$

or

(1) $$\Sigma(x - x_i)\sigma_i = 0$$

where

σ_1 is the area of x_2, x_3, x_4
σ_2 ,, ,, x_1, x_4, x_3
σ_3 ,, ,, x_4, x_1, x_2
σ_4 ,, ,, x_3, x_2, x_1

Let us write

$$b_i = (x - x_i)\sigma_i$$

so that the identity becomes

$$\Sigma b_i = 0$$

Here $b_1 + b_2 = 0$ is a point on the join of x_1 and x_2. But then $b_3 + b_4 = 0$, and the point is also on the join of x_3 and x_4. Thus $b_1 + b_2 = 0$ is a diagonal point.

Collectively the three diagonal points are:

$$(b_1 + b_2)(b_1 + b_3)(b_1 + b_4) = 0$$

or

$$\Sigma b_1 b_2 b_3 = 0$$

Two diagonal points are given by

$$(b_1 + b_3)(b_1 + b_4) = 0$$

or

$$b_1 b_2 = b_3 b_4$$

Hence

$$(b_1 b_2 - b_3 b_4)(b_1 b_3 - b_2 b_4)(b_1 b_4 - b_2 b_3)$$

gives the three points each twice and is to a numerical factor $(\Sigma b_1 b_2 b_3)^2$.

Exercise 21 – The numerical factor is -1.

The equation (1) implies, making x large, that

(2) $$\Sigma \sigma_i = 0$$

And also, writing ξ_i, η_i as rectangular co-ordinates of $x - x_i$, it implies

$$\Sigma(\xi_i + \iota\eta_i)\sigma_i = 0$$

that is

(3) $$\Sigma \xi_i \sigma_i = 0$$

where ξ_i are the distances from the four points to any line.

We may use these facts to name points and lines by homogeneous real numbers. We take a reference triangle

x_1, x_2, x_3, say of unit area; the areas σ_1, σ_2, σ_3 will name conveniently any finite point x; and numbers r_i proportional to them will equally serve.

Again, with the same triangle of reference, numbers ρ_i proportional to the distances from the vertices to any selected line will serve to name that line. Then we have from (3) the fundamental incidence relation

$$r_1\rho_1 + r_2\rho_2 + r_3\rho_3 = 0$$

expressing that the point r is on the line ρ.

The transition from the barycentric naming to the naming by a vector is then, writing in (1) $x=0$, $x_4 = x$,

$$(r_1 + r_2 + r_3)x = r_1x_1 + r_2x_2 + r_3x_3 \tag{4}$$

Given a triangle we wish to draw from the vertices to the opposite sides three segments of the same length λ, all being on one point.

Taking the point on the three segments as the base-point, let the vertices be x_1, x_2, x_3. And let the barycentric coordinates of x be r_1, r_2, r_3. Then the identity (1) is

$$\Sigma r_i(x - x_i) = x\Sigma r_i$$

and therefore a diagonal point is given by

$$r_1(x - x_1) = x\Sigma r_i$$

whence

$$x - x_1 = -\frac{x_1}{r_2 + r_3}\sum r_i$$

Hence

$$x_1\bar{x}_1/(r_2 + r_3)^2 = \lambda^2/(\Sigma r_i)^2$$

and hence, substituting for $\bar{x}_i$ in

$$r_1\bar{x}_1 + r_2\bar{x}_2 + r_3\bar{x}_3 = 0$$

$$\sum\frac{r_1(r_2 + r_3)^2}{x_1} = 0$$

$$2r_1r_2r_3(x_2x_3 + x_3x_1 + x_1x_2) + \Sigma r_2r_3x_1(r_2x_2 + r_3x_3)$$

or since

$$r_2x_2 + r_3x_3 = -r_1x_1$$

$$2(x_2x_3 + x_3x_1 + x_1x_2) - x_1^2 - x_2^2 - x_3^2 = 0$$

This is the rationalised form of

$$\sqrt{x_1} + \sqrt{x_2} + \sqrt{x_3} = 0$$

The sought point is therefore given by

$$\sum \sqrt{x - x_i} = 0$$

or

$$3x^2 - 2x(x_1 + x_2 + x_3) + 2\Sigma x_2 x_3 - \Sigma x_1{}^2 = 0$$

There are then two such points, say f and f'.

Exercise 22 – Prove that $3(x_1 - f)(x_1 - f') = -(x_2 - x_3)^2$, and hence that the two values of λ are the same.

Exercise 23 – The equation of the circumcircle in barycentric co-ordinates is

$$\Sigma(r_i t_i) \times (\Sigma r_i / t_i) = (\Sigma r_i)^2$$

or

$$p_{23} r_2 r_3 + p_{31} r_3 r_1 + p_{12} r_1 r_2 = 0$$

Exercise 24 – Hence the equation of the circumsphere of a reference tetrahedron is

$$\sum^{6} p_{12} r_1 r_2 = 0$$

§ **118. Foci** – There is a useful extension of the transition from the equation of a point to the vector of the point. Suppose a curve given by a real line-equation, a homogeneous algebraic equation with real coefficients in the co-ordinates ρ_i of a line, say

$$\phi(\rho_i) = 0$$

We take ρ_i as the distance from x_i to a variable line of the curve. If the line has a clinant t and if x is any point on it, then

$$2\iota \sqrt{t} \rho_i = x - x_i - t(\bar{x} - \bar{x}_i)$$

Thus the clinants of the lines of the curve which are on x are given by

(1) $$\phi\{x - x_i - t(\bar{x} - \bar{x}_i)\} = 0$$

When x is on the curve two values of t are equal. Thus the discriminant of (1) as to t will give the self-conjugate equation of the curve. More generally the discriminant as to z of

(2) $$\phi\{x - x_i - z(\bar{y} - \bar{x}_i)\}$$

will give the image-equation of the curve. Let (2), arranged in powers of z, be

$$c_0 + c_1 z + \dots + c_n z^n$$
$$= c_n(z - z_1) \dots (z - z_n)$$

The discriminant is to a numerical factor

$$c_n^{2(n-1)}\Pi(z_1 - z_2)^2$$

If then $c_0 = 0$, so that, say, $z_1 = 0$, it contains the factor $z_2^2 z_3^2 \ldots z_n^2$. It is then of the form

$$c_0 \mathrm{P} + c_1^2 \mathrm{Q}$$

Here

$$c_0 = \phi(x - x_i)$$

and c_1 is linear in $\bar{y}$.

Thus if $c_0 = 0$, then two values of $\bar{y}$ become equal.

Thus $\phi(x - x_i) = 0$ gives those points for which two images coincide. They are called branch-points of $f(x, \bar{y}) = 0$; but with respect to the curve $f(x, \bar{x}) = 0$ they are usually called the foci.

Thus the *foci of the curve* $\phi(\rho_i) = 0$ *are found by replacing* ρ_i *by* $x - x_i$. When the curve is a point, this is the above transition, § 117 (4).

If the relation (1) is not homogeneous we can render it so by the quadratic relation of § 38. If it contains the distances from any number of points the argument is unaltered. The result is that we make the substitution of $x - x_i$ for ρ_i in the highest powers alone.

For example, the equation

$$\rho_1 \rho_2 = \kappa$$

is a conic in lines. The foci are

$$(x - x_1)(x - x_2) = 0$$

And the foci of $\Sigma\lambda_i/\rho_i = 0$ are given by $\Sigma\lambda_i/(x - x_i) = 0$.

CHAPTER XVII

RATIONAL CURVES

§ 119. The Curves – We consider a rational algebraic function of z, of degree n, $x = (\alpha z)^n/(\beta z)^n$

(1) $x = (\alpha_0 + \alpha_1 z + \ldots + \alpha_n z^n)/(\beta_0 + \beta_1 z + \ldots + \beta_n z^n)$

or, in partial fractions,

(2) $$x = \Sigma c_i/(a_i - z)$$

when the constant α_n/β_n—the value of x for $z = \infty$ —has been made the base-point.

We now restrict z to the base-line, or to the base-circle. Then x moves on a curve R^n which we call a rational curve (algebraic being understood) or, when n is specified, a rational bi-n-ic curve. The integer n is the inversive degree of the curve. A rational curve being the rational map of a line or circle, is essentially an actual curve, a continuous closed succession of points of the plane. Thus, whereas the bilinear curve was a circle or an extra pair, the rational bilinear curve is a circle.

The number of real parameters in (1) is at first sight $4n + 2$. But the variable z is subject to a homography with three parameters. Thus the curve has $4n - 1$ parameters. There are three for the circle, seven for the rational biquadratic, eleven for the rational bicubic, and so on.

The transition between $x = (\alpha\rho)^n/(\beta\rho)^n$ and $x = (\alpha t)^n/(\beta t)^n$ is effected by any one-to-one correspondence between ρ and t, for instance, by

$$t = (1 + \iota\rho)/(1 - \iota\rho)$$

This does not alter the degree n.

An antigraphy applied to x, $\bar{y} = (ax + b)/(cx + d)$, gives for y again a partial fraction of degree n; so that a rational curve is an inversive thing. Under inversions it has $4n - 7$ real constants.

We are taking the rational fraction

(1) $$x=(\alpha z)^n/(\beta z)^n$$

and making z move on the base-line. But consider two points z, $\bar{z}$ images in the base-line. Let

$$y=(\alpha\bar{z})^n/(\beta\bar{z})^n$$

Then

(3) $$\bar{y}=(\bar{\alpha}z)^n/(\bar{\beta}z)^n$$

and if we eliminate z between (1) and (3) we have an equation

(4) $$f(x^n, \bar{y}^n)=0$$

If we eliminate $\bar{z}$ between the conjugate equations of (1) and (3) we have

$$f(y^n, \bar{x}^n)=0$$

Either is the image-equation associated with the curve. For any point x we have n images y. The curve itself is those points for which a point coincides with an image—that is, its self-conjugate equation is

$$f(x^n, \bar{x}^n)=0$$

§ **120. The Foci** – We define the foci of the curve as those points x for which two values of z are equal. They are given by the discriminant of

$$x(\beta z)^n-(\alpha z)^n$$

The discriminant of a polynomial is of degree $2(n-1)$ in the coefficients. There are then in general $2(n-1)$ foci.

The values of z which give the foci are the zeros of the Jacobian of $(\alpha z)^n$ and $(\beta z)^n$—that is, in the abridged notation $|\,\alpha\beta\,|\,(\alpha z)^{n-1}(\beta z)^{n-1}$. They are given in practice by $dx/dz=0$.

Where x is a focus, two values of z become equal, two values of $\bar{z}$ become equal, and therefore two images of x coincide, at a double image.

Consider the case

$$2x=z+\iota/z$$

The discriminant of $2xz=z^2+\iota$ gives $x^2=\iota$.

The equal values of z are given by $z^2=\iota$. To $z=e^{\iota\pi/4}$, corresponds the focus $x=e^{\iota\pi/4}$. To $z=e^{-\iota\pi/4}$, corresponds $2y=e^{-\iota\pi/4}+e^{\iota\pi/2+\iota\pi/4}=0$.

The image-equation is, since $2x = \rho + \iota/\rho$, $2\bar{y} = \rho - \iota/\rho$,

$$x^2 - \bar{y}^2 = \iota$$

From this we see again that the two values of y are equal when $x^2 = \iota$.

Thus for the rectangular hyperbola

$$2x = \rho + \iota/\rho$$

there are two foci. Each has a double image, and here the double images coincide.

Instead of foci and their double images we may speak of branch-points and their crossing-points, with reference to the equation $f(x^n, \bar{y}^n) = 0$.

§ **121. The Double Points** – When $x = (\alpha z)^n/(\beta z)^n$ there are for a given x n points z, say $z_1, z_2 \ldots z_n$. The relation of these is found from

$$(\alpha z_1)^n/(\beta z_1)^n = (\alpha z_2)^n/(\beta z_2)^n$$

On removing the factor $z_1 - z_2$ we have a symmetrical equation, say $Az_1{}^{n-1}z_2{}^{n-1} = 0$, of degree $n - 1$ in each. For the full study of this we should develop it in a Clebsch series; the first term is when $z_1 = z_2$, giving the Jacobian of $(\alpha z)^n$ and $(\beta z)^n$.

It is sufficient here to notice that when z_1 is ρ, z_2 (that is, all the other z's) describes a residual curve which is found by elimination of ρ from $Az^{n-1}\rho^{n-1}$ and $\bar{A}\bar{z}^{n-1}\rho^{n-1}$.

By the rule of elimination we get for the residual curve

$$A^{n-1}\bar{A}^{n-1}z^{(n-1)^2}\bar{z}^{(n-1)^2}$$

It has therefore with the axis of reals $2(n-1)^2$ common points, intersections or common image-pairs.

Now the intersections pair off, for if $z_1 = \rho_1$ gives $z = \rho_2$, then $z_1 = \rho_2$ gives $z = \rho_1$. There are then $(n-1)^2$ pairs of intersections and image-pairs.

Therefore the rational curve R^n has $(n-1)^2$ special points which correspond either to two values of ρ or to two conjugate values. These we call double points (or nodes). When the values are real, the curve intersects itself, and the double point may be called actual or hyperbolic. In the other case, the double point is isolated, or elliptic.

A double point may be regarded as the union of two foci. Thus the general algebraic curve, $f(x^n, \bar{x}^n)$, has $2n(n-1)$ foci. But when it is rational it has $2(n-1)$ foci and $(n-1)^2$ double points.

And

$$2(n-1)+2(n-1)^2=2n(n-1)$$

§ 122. **Cusps** – A focus may coincide with its double image. Then we have a cusp, or stationary point. The focus is given by $dx/dz=0$; that is, by the Jacobian of $(\alpha z)^n$ and $(\beta z)^n$. For a cusp, the variable must take one of these $2(n-1)$ values. When z is a real, then $dz/d\rho$ must vanish for a real; and when z is a turn, then dz/dt must vanish for a turn.

Thus when $x=(\alpha\rho)^n/(\beta\rho)^n$ the foci are given by

$$|\,\alpha\beta\,|\,(\alpha\rho)^{n-1}(\beta\rho)^{n-1}$$

If between this and its conjugate we eliminate ρ, we have the condition for a cusp, of degree $2(n-1)$ in α_i, β_i, $\bar{\alpha}_i$, $\bar{\beta}_i$.

A cusp is on the curve; but it is not correct to say that when a focus is on the curve we have a cusp. To a focus correspond z_1, $z_2=z_1$, z_3 . . . z_n. If the base-line is on z_3, we have a focus on the curve, but no cusp.

A further singularity is when y is the double image of x, and x is the double image of y.

§ 123. **The Curve R^2** – In this case, $x=(\alpha z)^2/(\beta z)^2$, where z moves on a circle. There are two foci, f_1 and f_2, for each of which the values of z are equal. Thus the equation can be written

$$(x-f_1)/(x-f_2)=\kappa(z-z_1)^2/(z-z_2)^2$$

or, by independent homographies on x and z,

$$2x=z^2 \tag{1}$$

This then is a canonical form for the 1 : 2 correspondence. When z describes a circle, x describes an R^2 with foci at 0 and ∞. This is in Euclidean geometry a limaçon. To a given x correspond z and $-z$, and as z describes its circle, $-z$ describes an equal circle. If these circles intersect, the curve R^2 has an actual double point. If they do not intersect, their common image-pair gives the curve an isolated double point. If they touch, the curve has a cusp. But they can touch only at 0 or ∞. In the former case we have in Euclidean geometry a cardioid, in the latter a parabola. Thus the parabola is a curve given by $2x = z^2$, where z describes a line. It has the focus 0, and the cusp ∞.

When there is a double point, let us place it at ∞, and take as the foci $x = \pm 1$, corresponding to $z = \pm 1$. The correspondence is then, if to $z = \infty$ corresponds $x = \infty$,

$$(x-1)/(x+1) = (z-1)^2/(z+1)^2$$

or

$$2x = z + 1/z \tag{2}$$

The parameters of the double point are $z = 0$, $z = \infty$. The circles are either on these points or about these points. If we write then $z = \rho t$, we have for $\rho =$ constant an R^2 with an isolated double point at ∞, the ellipse in Euclidean geometry, and for $t =$ constant an R^2 with an actual double point at ∞, the hyperbola.

We shall discuss these curves in the next chapter.

§ 124. **Sections of R^n by a Circle** – We denote the general algebraic curve

$$x^n \bar{x}^n + \ . \ . \ .$$

by C^n. In general two curves C^m and C^n have $2mn$ common points. If C^m is on a double point of C^n, this counts for two common points. If C^m and C^n have a common double point this counts for four common points.

The bilinear curve C and R^n have then $2n$ common points. Let the parameters of these be ρ_i. Since three of these points determine the C, there is between any four of the ρ_i a symmetric relation

$$S(\rho_1{}^{2n-3}, \quad \rho_2{}^{2n-3}, \quad \rho_3{}^{2n-3}, \quad \rho_4{}^{2n-3}) = 0$$

If $\rho_1 = \rho_2$, the C is tangent; if $\rho_1 = \rho_3$, and $\rho_2 = \rho_4$, it is bitangent.

Thus a bitangent C is given by

$$S(\rho_1{}^{4n-6}, \qquad \rho_2{}^{4n-6}) = 0$$

and we may expect $4n - 6$ systems of such bitangents.

If $\rho_1 = \rho_2 = \rho_3$, the C is osculating. If $\rho_1 = \rho_2 = \rho_3 = \rho_4$, the C is stationary. We have then an equation of degree

$4(2n-3)$ in ρ_1. This gives the parameters of the apses. For a real ρ_1 the apse is actual, and when C is a circle it meets R^n there at four consecutive points. Conjugate values of ρ_1 give pairs of isolated apses.

There are then $4(2n-3)$ apses.

§ 125. Sections of an R^n by a C^{n-2} – Let us cut the curve R^n by a curve C^{n-2}. We are to substitute in

$$C^{n-2} \equiv x^{n-2}\bar{x}^{n-2} + \ . \ . \ .$$
$$x = (\alpha\rho)^n/(\beta\rho)^n \quad \text{and} \quad \bar{x} = (\bar{\alpha}\rho)^n/(\bar{\beta}\rho)^n$$

If the $2n(n-2)$ intersections be ρ_i, where $i = 1 \ . \ . \ . \ 2n(n-2)$, we have

$$C^{n-2} \equiv \alpha_0{}^{n-2}\bar{\alpha}_0{}^{n-2}\Pi(\rho-\rho_i)/(\beta\rho)^n(\bar{\beta}\rho)^n$$

where x is any point of the curve whose parameter is ρ. Let x be an actual double point, with the two parameters a_1, b_1. Then

$$\Pi(a_1-\rho_i)/(\beta a_1)^n(\bar{\beta}a_1)^n = \text{same in } b_1$$

Therefore $\Pi(a_1-\rho_i)/\Pi(b_1-\rho_i)$ is a constant κ_1 for all curves C^{n-2}.

To determine the constant, consider the curve C^{n-2} on the $n(n-2)$ other double points. Let the parameters of a double point be a_i, b_i. Then

$$\kappa_1 = \frac{a_1-a_2 \ . \ a_1-a_3 \ . \ . \ . \ a_1-b_2 \ . \ a_1-b_3 \ . \ . \ .}{b_1-a_2 \ . \ b_1-a_3 \ . \ . \ . \ b_1-b_2 \ . \ b_1-b_3 \ . \ . \ .}$$

or if

$$\phi(x) = \Pi(x-a_i)(x-b_i)$$
$$\kappa_1 = -\phi'(a_1)/\phi'(b_1)$$

Thus we have for the parameters ρ_i the $(n-1)^2$ equations

$$(1) \qquad \Pi(a_j-\rho_i)/\phi'(a_j) + \Pi(b_j-\rho_i)/\phi'(b_j) = 0$$

But, since by the theory of partial fractions the sum of these expressions is 0, there are $n(n-2)$ independent equations on the $2n(n-2)$ numbers ρ_i.

Hence there are curves C^{n-2} of contact, which are tangent to R^n n times. For if we make the ρ_i two-by-two equal, we have $n(n-2)$ equations in $n(n-2)$ numbers. On these contact curves, see F. Morley and W. K. Morrill, *American Journal*, vol. 54, 1932.

If we cut the curve R^n by a $C^{2(n-2)}$ we have by similar argument for the $4n(n-2)$ intersections,

$$\Pi(a_j - \rho_i)/\Pi(b_j - \rho_l) = \kappa_j'$$

To determine the constant we take the curve C^{n-2} on the other double points twice, so that $\kappa_j' = \kappa_j^2$.

We have then the $(n-1)^2$ equations

$$(2) \qquad \Pi(a_j - \rho_i)/\phi'^2(a_j) - \Pi(b_j - \rho_i)/\phi'^2(b_j) = 0$$

on the $4n(n-2)$ parameters, these equations being now independent.

Thus when the curve is an R^3, a curve C^2 meets it in twelve points, subject to the four conditions (2). And eight points determine a C^2.

When the curve is an R^4, a curve C^4 meets it in thirty-two points, subject to nine conditions. And twenty-three points on the given R^4 serve to determine the pencil $R^4 + \lambda C^4$. When the curve is an R^5, a curve C^6 meets it in sixty points, subject to sixteen conditions. And forty-four points on the given R^5 serve to determine the pencil $R^5C + \lambda C^6$, C being any bilinear curve.

§ **126. Degeneration** – A curve given by a self-conjugate equation $f(x, \bar{x}) = 0$ breaks up when f is a product, $f = f_1 f_2$.

Let us consider one way in which a curve given by a map-equation can break up. Suppose that

$$x = (a_0 + a_1\rho + \ . \ . \ .)/(\beta_0 + \beta_1\rho + \ . \ . \ .)$$

Let $x = 0$ correspond to $\rho = 0$, so that $a_0 = 0$, and

$$(1) \qquad x = (a_1\rho + \ . \ . \ .)/(\beta_0 + \beta_1\rho + \ . \ . \ .)$$

If now $\beta_0 = 0$ the curve breaks up into

$$x = (a_1 + a_2\rho + \ . \ . \ .)/(\beta_1 + \beta_2\rho + \ . \ . \ .)$$

and a something. To find that something, make in (1) β_0 real by dividing numerator and denominator by its direction, and then write $\rho = \beta_0\rho'$. Then (1) is

$$x = \beta_0\rho'(a_1 + a_2\beta_0\rho' + \ . \ . \ .)/(\beta_0 + \beta_1\beta_0\rho' + \ . \ . \ .)$$

with the limit

$$x = a_1\rho'/(1 + \beta_1\rho')$$

which is a circle.

Thus an R^2 can degenerate into two circles; an R^3 can degenerate into an R^2 and a circle, having five double points, or into three circles, having six double points.

From n circles, mutually intersecting, an idea of the look

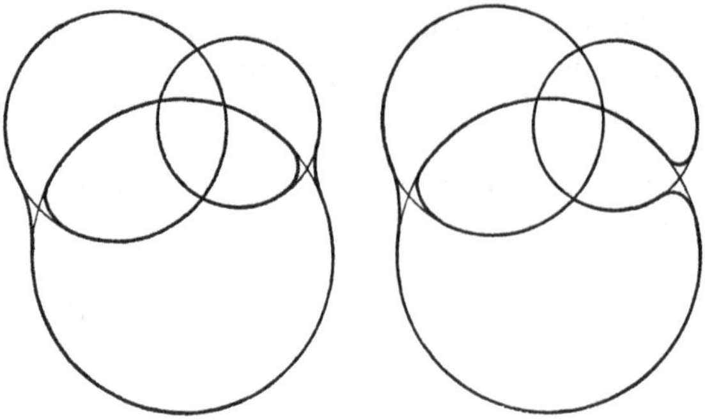

FIG. 63

of one type of R^n is easily formed. Take the case of three circles, and round off two of the intersections in such a way that a continuous curve is formed; as in fig. 63.

§ 127. Mechanical Description of R^n – If the curve be given by

$$(1) \qquad y=\sum_1^n \frac{A_i}{a_i-t}$$

we write $(a_n-t)(\bar{a}_n-1/\tau)=a_n\bar{a}_n-1$.

Expressed by proper fractions in τ, y becomes

$$y=B_0+B_1/\tau+\sum_1^{n-1}\frac{B_i}{b_i-\tau}$$

Moving the base-point to B_0, and replacing B_1 by x, we have the equation of a motion in the form

$$y=x/\tau+\sum_1^{n-1}\frac{B_i}{b_i-\tau}$$

When $x=0$, the slot is

$$y=\sum_1^{n-1}\frac{B_i}{b_i-\tau}$$

and when $y=0$, the slot is

$$x=\sum_1^{n-1}\frac{B_i}{1-b_i/\tau}$$

Repeating the process for each slot, we come to the case of two circular slots, or three-bar motion (rational).

There remains the more general case when in the given expression of y in partial fractions there are repeated poles, as, for instance, a term $A/(a-t)^2$. Geometrically speaking, the curve has then a focus at infinity. The easy way is to allow inversions. After a suitable inversion, any rational curve is expressed in the form (1), the foci being now all finite.

Thus the general rational curve, after an inversion when necessary, can be mechanically described by two pins moving in two slots, the slots being simpler than the proposed curve.

CHAPTER XVIII

CONICS

§ 128. **The Parabola** – The parabola, analytically, may be defined by

(1) $$2x = z^2$$

where z describes a line. We suppose x and z to be points of the same plane. We take for the line the vertical at the point 1. Its equation is either $z = 1 + \iota\rho$ or $z = 2/(1-t)$, so that the curve is, for a real parameter,

(2) $$2x = (1 + \iota\rho)^2$$

or for a turn

(3) $$x = 2/(1-t)^2$$

If $x = \xi + \iota\eta$, then in (2)

$$2\xi = 1 - \rho^2, \qquad \eta = \rho$$

Thus the selected real parameter is the *ordinate* of x.

Hence

$$\xi^2 + \eta^2 = (1 + \rho^2)^2/4 = (1 - \xi)^2$$

that is, the distance from the point 0 is the distance from the vertical line. This is the usual defining property.

The focus is $x = 0$, the directrix is $x = 1 + \iota\rho$, the apse or vertex is $x = 1/2$.

§ 129. **Lines of the Curve** – If we polarise the above (2) or (3) we have

(1) $$2x = (1 + \iota\rho)(1 + \iota\rho_1)$$

(2) $$x = 2/(1-t)(1-t_1)$$

In this second case, we clear of fractions before polarising. Taking (1) for given ρ_1 we have a line, and for given ρ we have a line. By the theory of envelopes, all such lines touch the curve for which $dx/d\rho$ has the same clinant as $dx/d\rho_1$.

This requires that $\rho = \rho_1$. The lines then are the tangents of the parabola, or simply its lines. Similarly, for (2), $t dx/dt$ and $t_1 dx/dt_1$ must have the same clinant (§ 84), whence for the envelope $t = t_1$. But this does not require independent proof; (2) is merely the transform of (1) by $1 + \iota\rho = 2/(1 - t)$.

Since in (1) $x = 0$ when $\rho = \iota$, the image of the focus in the tangent at ρ_1 is when $\rho = -\iota$, that is,

$$x = 1 + \iota\rho_1$$

The parabola is then the locus of a line such that the image in it of a fixed point is on a fixed line.

Exercise 1 – The normal, from the axis to the point ϱ, is $1 + \iota\varrho$.

We have in (1) or (2) the expression for any point outside the parabola, in terms of the parameters of the two tangents from it. Notice that in general there is a profound difference between the regions in which foci of a curve lie, and the other regions. We have here for such outside points—osculant points—

$$2x = (1 + \iota\rho_1)(1 + \iota\rho_2)$$

or

$$x = 2/(1 - t_1)(1 - t_2)$$

We shall proceed with the latter.

§ **130. Theory of the Four-line** – Consider three lines t_i of the parabola. The join of t_1, t_2 is

$$x = 2/(1 - t_1)(1 - t_2)$$

Now the equation

(1) $$x = 2(1 - t)/(1 - t_1)(1 - t_2)(1 = t_3)$$

gives a circle, and when $t = t_3$ x is the join of t_1, t_2. Thus (1) is the circumcircle of the three-line t_i. When $t = 1$, $x = 0$; thus the circumcircle of any three lines of a parabola is on the focus.

The centre of the circle (1)—that is, the circumcentre of the three lines—is

(2) $$x = 2/(1 - t_1)(1 - t_2)(1 - t_3)$$

And the equation

(3) $$x = 2(1 - t)/(1 - t_1)(1 - t_2)(1 - t_3)(1 - t_4)$$

is that of a circle, which for $t=t_4$ is on the point (2). Hence we have Wallace's theorem that the circumcentres of all three lines selected from four (which for brevity we will call 3/4 lines) are on a circle, which is also on the focus of the parabola which touches the four lines—that is, on the point at which the four circumcircles intersect.

We call this circle (3) the centric circle of the four lines, and its centre

$$x=2/(1-t_1)(1-t_2)(1-t_3)(1-t_4)$$

the centric point, or simply the centric.

Evidently we have a chain of theorems which can be continued indefinitely. But when we take more than four lines, as tangents of a parabola, they are not general. It is for four lines that the parabola gives a canonical treatment.

Exercise 2 – The orthocentre of three lines of a parabola is on the directrix.

Exercise 3 – Hence the orthocentres of 3/4 lines are on a line.

Exercise 4 – The mid-points of the three diagonals of a four-line are on a line, and this line is perpendicular to the directrix.

Denote the join of the lines t_1, t_2 by x_{12}, and so forth. Then

$$x_{12}=2/(1-t_1)(1-t_2)$$
$$x_{34}=2/(1-t_3)(1-t_4)$$

Hence

$$x_{12}x_{34}=x_{13}x_{24}=x_{14}x_{23}=4/\Pi_4$$

where

$$\Pi_4=(1-t_1)(1-t_2)(1-t_3)(1-t_4)$$

Thus the three pairs of points are in the pairing or involution

$$xy=4/\Pi_4 \tag{4}$$

Regarding (4) as a homography (of period 2) it sends the line t_1, on which are x_{12}, x_{13}, x_{14}, into the circle on x_{23}, x_{34}, x_{42}; and so for t_i. It sends ∞ into the focus 0. And it sends the centric circle into the directrix.

A four-line may thus be completed into a configuration of eight points and eight circles, four points on each circle and four circles on each point. Let the four lines, or circles on ∞, be 1, 2, 3, 4. They give six points, 12, . . . These give four circles, 1 2 3; and these are on a point 1 2 3 4. On the point 12 are the circles 1, 2, 123, 124. On the circle 123 are the points 12, 23, 31, 1234.

That three pairs of a quadratic involution determine a fourth in a mutual manner is a case of the proposition that any three quadratics determine a fourth in a mutual manner (§ 57), there being a linear identity between the four squares.

Exercise 5 – The quadratics being here

$$\begin{aligned} x^2+a_1x+4/\Pi_4&=0\\ x^2+a_2x+4/\Pi_4&=0\\ x^2+a_3x+4/\Pi_4&=0\\ x&=0 \end{aligned}$$

the identity is

$$\Sigma(a_2-a_3)(x^2+a_1x+1/\Pi_4)^2+(a_2-a_3)(a_3-a_1)(a_1-a_2)x^2\equiv 0$$

§ 131. Sections by a Circle – Let us here use the parameter ρ. The section of

$$(x-x_0)(\bar{x}-\bar{x}_0)+\kappa=0$$

by

$$2x=(1+\iota\rho)^2$$

is the quartic

$$(1+\rho^2)^2-2\bar{x}_0(1+\iota\rho)^2-2x_0(1-\iota\rho)^2+4(x_0\bar{x}_0+\kappa)=0$$

whence

(1)
$$\begin{aligned} s_1&=\rho_1+\rho_2+\rho_3+\rho_4=0\\ s_2&=2(1+x_0+\bar{x}_0)\\ s_3&=4\iota(\bar{x}_0-x_0)\\ s_4&=1-2(x_0+\bar{x}_0)+4(x_0\bar{x}_0+\kappa) \end{aligned}$$

When $\rho_1=\rho_2=\rho_3=\rho_4$ we have an apse. There is then but one apse, $\rho=0$.

Exercise 6 – The stationary circle, meeting four times at the apse, is the circle $(-1/2, 1)$.

When $\rho_1=\rho_2=\rho_3=\rho$, then $\rho_4=-3\rho$. This gives the circle of curvature at the point ρ.

When $\rho_1=\rho_3$, $\rho_2=\rho_4$ we have the bitangent circles. Then

$$\begin{aligned} \rho_1+\rho_2&=0\\ \rho_1^2+\rho_2^2+4\rho_1\rho_2&=2(1+x_0+\bar{x}_0)\\ x_0=\bar{x}_0&=\text{a real, } \xi_0\\ \rho_1^2\rho_2^2&=1-2(x_0+\bar{x}_0)+4(x_0\bar{x}_0+\kappa) \end{aligned}$$

that is

$$2\xi_0 = -1 - \rho_1{}^2$$

and

$$\kappa = -1 - \rho_1{}^2$$

For real values of ρ_1 we have circles with actual contact, ending with the stationary circle $\rho_1 = 0$. For $-1 < \rho_1{}^2 < 0$ we have real circles having with the parabola a double image-pair. For $\rho_1{}^2 = -1$ we have the focus. And for $-1 > \rho_1{}^2$ the circles become extra pairs. Writing $\kappa = \zeta^2$, the locus of extra pairs is

$$\zeta^2 = 2\xi$$

a parabola in a vertical plane, with its apse at the focus of the given parabola, and with its focus at the apse. The relation of the two parabolas is mutual; each is the locus of imaginary circles (that is, extra pairs) bitangent to the other.

Exercise 7 – The system of bitangent circles may be written

$$x = 2\lambda^2 + \lambda t$$

when λ is real. Verify that the envelope is a parabola.

§ 132. **Bifocal Conics** – The general R^2 is a rational curve with a double point, two foci, and four apses.

We take the double point at ∞, and the equation

(1) $$2x = z + 1/z$$

where $z = \rho t$ and ρ is positive. The foci are the points $x = \pm 1$, for which the quadratic in z has equal roots, either 1 or -1.

The curves for which $\rho =$ constant we call ellipses. They have an isolated double point at ∞, for which $t = 0$ or ∞.

The curves for which $t =$ constant we call hyperbolas. They have an actual double point at ∞, for which $\rho = 0$ or ∞. The curve being

$$2x = \rho t_0 + 1/\rho t_0$$

the approximations

$$\begin{aligned} 2x &= \rho t_0 \qquad \text{for } \rho \text{ large} \\ 2x &= 1/\rho t_0 \qquad \text{for } \rho \text{ small} \end{aligned}$$

give the asymptotes.

The curves are confocal conics, the map by (1) of the

orthogonal system of concentric circles and their rays. Let us handle them simultaneously. We have from (1)

$$2(x-1)=(z-1)^2/z$$
$$2(\bar{x}-1)=(\bar{z}-1)^2/\bar{z}$$

When $z=\rho t$, $\bar{z}=\rho/t$, we have then

$$2\mid x-1\mid=(z-1)(\bar{z}-1)/\rho$$
$$=\rho+1/\rho-t-1/t$$

For $\rho+1/\rho\geqslant 2$, while $t+1/t\leqslant 2$.

Hence the sum of the focal distances is $\rho+1/\rho$, and the difference is $\pm(t+1/t)$; the former is constant for the ellipse, the latter for the hyperbola. The differential form of (1) is

(2) $$dx/\sqrt{x^2-1}=dz/z=d\rho/\rho+\iota d\theta$$

Thus for the ellipse $dx/\sqrt{x^2-1}=\iota d\theta$, so that the direction of dx is that of the exterior bisector of the angle -1, x, 1. Whereas for the hyperbola $dx/\sqrt{x^2-1}=d\rho/\rho$, and the direction of dx is that of the interior bisector. The two curves on a point cut then at right angles, as we knew from the mapping.

§ 133. **Lines of the Curve** – The polarised form of (1) above is

(1) $$x(z+z_1)=zz_1+1$$

Thus for the ellipse we have

(2) $$x=(\rho^2tt_1+1)/(t+t_1)$$

which for given t_1 is a line of the curve. Thus (2) is the intersection of the lines t and t_1. It is an osculant point, outside the curve.

The image of the focus 1 in the line t_1 is the map of the image of the point $z=1$ in the circle $\mid z\mid=\rho$—that is, of $z=\rho^2$. Thus the image of the focus 1 in the line t_1 is, writing $\rho t=\rho^2$ or $t=\rho$,

$$x=(\rho^3t_1+1)/\rho(\rho+t_1)$$

or

(3) $$x-1=(\rho t_1-1)(\rho-1/\rho)/(\rho+t_1)$$

Similarly the image of the focus -1 is given by

$$x'-1=(\rho t_1+1)(\rho-1/\rho)/(\rho-t_1)$$

Hence

$$(\bar{x} - 1)(x' - 1) = (\rho - 1/\rho)^2$$

that is, the product of distances from the foci to any tangent is a constant.

The argument for the hyperbola is similar, (2) being replaced by

$$x = (t^2\rho\rho_1 + 1)/t(\rho + \rho_1)$$

Exercise 8 – For varying t_1, (3) is a circle. Prove that ∞^1 triangles can be inscribed in it and circumscribed to the ellipse, and that the orthocentre of any such triangle is the focus -1.

§ **134. Theory of the Five-line** – Let us write the polarised form (1, § 133) as

$$(x + 1)/(x - 1) = (z + 1)(z_1 + 1)/(z - 1)(z_1 - 1)$$

and write u_i for $(z_i + 1)/(z_i - 1)$.

Then for the ellipse, where $z_i = \rho t_i$, the osculant point or join of the lines t_1 and t_2 is

(1) $$(x + 1)/(x - 1) = u_1 u_2$$

Call this point x (12). Consider

(2) $$(x + 1)/(x - 1) = u_1 u_2 u_3 u_4$$

This is for varying t_4 a circle. When t_4 is $-t_3$, it is the point (12). Thus the circle is on (12) and similarly it is on (23) and (31). It is then the circumcircle of the triangle (123) formed by the three tangents. Thus the equation (2) restates the theorem that the circumcircles of 3/4 lines meet at a point, the focus of the parabola on the four-lines. It is convenient—though unfair to Wallace and Miquel—to call this the Clifford point of the four-lines. Calling it (1234), we write

(3) $$(x + 1)/(x - 1) = u_1 u_2 u_3 u_4 u_5 u_6$$

This for varying t_6 is a circle. When t_6 is $-t_5$, it becomes (1234). Equally it is on the points (2345), (3451), (4512), (5123). This is Miquel's theorem that the foci of the parabolas on 4/5 lines are on a circle—the Miquel circle of a five-line.

Also when six lines touch an ellipse, the six Miquel circles meet at a point given by (3)—the Clifford point of the six-line.

The above argument proves the same theorems for the case when the five lines touch a hyperbola, ρ being substituted for t.

We thus have the complete five-line. We take the point ∞ and the lines 1, 2, 3, 4, 5. The points 12, 23, 31 are on the circle 123. The circles 123, 234, 345, 412 are on the point 1234. The five points such as 1234 are on the circle 12345.

Inversively this is a Clifford configuration Γ_5 of sixteen points and sixteen circles; on each circle are five points, and on each point are five circles.

And when six lines touch a conic we have a special complete six-line. We take the point ∞ and the six lines 1, 2, 3, 4, 5, 6. The six circles, such as 12345, meet now at the point 123456. Inversively this is a special Clifford configuration Γ_6 of thirty-two points and thirty-two circles, the points being

$$\infty, \qquad 12\ldots, \qquad 1234\ldots, \qquad 123456$$

and the circles

$$1\ldots 6, \qquad 123\ldots, \qquad 12345\ldots$$

On each circle are six points, and on each point are six circles.

This configuration is sent into itself by the involution

$$(4) \qquad (x+1)(x'+1)/(x-1)(x'-1) = u_1u_2u_3u_4u_5u_6$$

Here ∞ corresponds to the Clifford point (3); the point 12 corresponds to the point 3456; the line 1 corresponds to the circle 23456, and the circle 123 to the circle 456.

Also the focus $x=1$ corresponds to the focus $x'=-1$. To the conic itself corresponds an R^2 touching the six Miquel circles.

A special case is when a fixed point of (4) is ∞. The other is then 0, the centre of the conic; and the involution is merely $x+x'=0$.

§ 135. **Section by a Circle** – The circle $x\bar{x} - \bar{c}x - c\bar{x} + \kappa = 0$ meets the ellipse $2x = \rho t + 1/\rho t$ when

$$(\rho t + 1/\rho t)(\rho/t + t/\rho) + \ldots = 0$$

or

$$(1) \qquad t^4 - 2\bar{c}(\rho t^3 + t/\rho) - 2c(\rho t + t^3/\rho) + t^2(\rho^2 + 1/\rho^2 + 4\kappa) + 1 = 0$$

Hence

(2) $$s_4 \equiv t_1 t_2 t_3 t_4 = 1$$

or, in terms of θ_i when $t_i = e^{\iota\theta_i}$,

$$\Sigma\theta_i = 0 \quad (\text{mod } 2\pi)$$

There are then four apses given by $t_1 = t_2 = t_3 = t_4$, that is by

$$t^4 - 1 = 0$$

and sections by a circle give parameters apolar to this quartic.

The osculating circles are determined by $t_1{}^3 t = 1$

Exercise 9 – Through a point t of the ellipse can be drawn three circles which osculate it elsewhere ; and the circle on the points of osculation is on t.

The bitangent circles are determined by $t_1{}^2 t^2 = 1$; that is, by $t_1 t_2 = \pm 1$. For $t_1 t_2 = 1$ the centres lie on the focal axis ; for $t_1 t_2 = -1$ they lie on the transverse axis. We have from (1)

$$s_1 = 2(t_1 + t_2) = 2(\bar{c}\rho + c/\rho)$$
$$s_2 = t_1{}^2 + t_2{}^2 + 4t_1 t_2 = \rho^2 + 1/\rho^2 + 4\kappa$$
$$s_3 = 2t_1 t_2(t_1 + t_2) = 2(\bar{c}/\rho + c\rho)$$

Taking the system $t_1 t_2 = 1$, we have

$$t_1 + t_2 = \bar{c}\rho + c/\rho = \bar{c}/\rho + c\rho$$

whence $c = \bar{c} = \xi$ say, and

$$\xi^2(\rho + 1/\rho)^2 + 2 = \rho^2 + 1/\rho^2 + 4\kappa$$

When the circle is real, of radius r, then $\kappa = c\bar{c} - r^2 = \xi^2 - r^2$ so that the relation of the centre ξ and the radius is

(3) $$(\xi^2 - 1)(\rho - 1/\rho)^2 + 4r^2 = 0$$

When the circle is imaginary—that is, points ξ, $\pm\zeta$ in the vertical plane through the foci, then

(4) $$(\xi^2 - 1)(\rho - 1/\rho)^2 = 4\zeta^2$$

This is a hyperbola in the vertical plane, with apses at the foci $\xi = \pm 1$ of the ellipse, and with foci at the apses $\xi = \pm\frac{1}{2}(\rho + 1/\rho)$ of the ellipse. The reciprocity of the two conics is to be noticed, as in § 131.

For the hyperbola, in the argument ρ and t have to be interchanged. The apses are now given by

$$\rho^4 - 1 = 0$$

so that only two are actual. For the focal system of bitangent circles, the imaginary circles are now points on an ellipse in the vertical plane, as required in the above reciprocity.

This reciprocity exists also for the general C^2. Here there are four foci, either concyclic or anticyclic. Taking the former case, with each focus goes an apse. We consider these four out of the sixteen apses where the four circles of the inversions which send the curve into itself meet the curve. The imaginary bitangent circles of the focal system fill a second C^2 on a sphere, apses and foci being interchanged.

§ **136. The Images of a Point** – The image-equation of a conic with foci ± 1 is

$$x^2 + \bar{y}^2 + 2\mu x\bar{y} + \mu^2 = 1 \qquad (1)$$

For the discriminant as to $\bar{y}$ is $(1 - \mu^2)(x^2 - 1)$.

To determine the two images, x_1 and x_{-1}, of a point x_0 we take the equation $2x = z + 1/z$.

Let the given curve be an ellipse, the map of $z = \rho_0 t$. It is the map of the circle $|z| = \rho_0$, but it is equally the map of the circle $|z| = 1/\rho_0$. The images of any point z_0 in these two circles, $z_1 = \rho_0{}^2/\bar{z}_0$ and $z_{-1} = 1/\rho_0{}^2\bar{z}$, map into the points y_1 and y_2.

Thus:

$$2x = \rho t + 1/\rho t$$

$$2x_1 = \rho_0{}^2 t/\rho + \rho/\rho_0{}^2 t$$

$$2(x - x_1) = (\rho^2 - \rho_0{}^2)(t - 1/\rho_0{}^2 t)/\rho \qquad (2)$$

Since the points z, z_1, z_{-1} are on the same ray the points x, x_1, x_{-1} are on the same branch of a confocal hyperbola. From (2) the direction of $x - x_1$ depends solely on t; it is the same for all points x on this branch, and in particular when x is on both the ellipse and the branch it is the direction of the tangent to the branch (the normal to the ellipse).

The construction (due to Laguerre *) of the images of x_0 as

* E. Study, *Vorlesungen über Ausgewählte Gegenstände der Geometrie*, vol. 1, p. 101.

to an ellipse is this. Draw the branch on x of the confocal hyperbola, and draw from x the parallels to the tangents where this branch meets the ellipse. These give on the branch the images of x.

Similarly for the images of a point in the hyperbola or the parabola.

From x and x_1 the whole two-way sequence is determined uniquely. The Green function for the point x_0—that is, the simplest function with zeros at x_0, $x_{\pm 2}$, $x_{\pm 4}$ . . . and poles at $x_{\pm 1}$, $x_{\pm 2}$ . . ., follows with the use of elliptic functions.

Exercise 10 – For the parabola $\sqrt{x}+\sqrt{\bar{x}}=1$ the focus has the double image $x=1$. This has the new image 4. The sequence of images beginning with the focus is thus 0, 1, 4, 9 . . . n^2. . .

Exercise 11 – There are four normals from a point to a conic. The feet of these form a special four-point, being determined by 7 conditions (5 for the conic and 2 for the point). Show that the invariant which vanishes is

$$3I_1^2I_3+4I_2I_3'$$

where

$$I_1=|x_i\bar{x}_i,\ x_i,\ \bar{x}_i,\ 1|$$
$$I_2=\sum^3\lambda_{23}\lambda_{14}(\lambda_{13}+\lambda_{24}-\lambda_{12}-\lambda_{34})$$
$$I_3=\prod^3(\lambda_{13}+\lambda_{24}-\lambda_{12}-\lambda_{34})$$
$$I_4=|x_i^2,\ x_i,\ \bar{x}_i,\ 1|\times\text{conjugate}$$

Exercise 12 – Here I_1^2 is to a numerical factor Cayley's invariant

$$|\lambda_{i1}\quad\lambda_{i2}\quad\lambda_{i3}\quad\lambda_{i4}|$$

And

$$I_3'-I_1^2=\sum(x_1\bar{x}_2-x_2\bar{x}_1)^2\sigma_1\sigma_2$$

where

$$\sigma_1=\begin{vmatrix} x_2 & \bar{x}_2 & 1 \\ x_3 & \bar{x}_3 & 1 \\ x_4 & \bar{x}_4 & 1 \end{vmatrix}$$

Hence (taking x_1 to be 0) $I_3'-I_1^2$ is to a numerical factor the product of the areas of the four triangles. Hence the relation on the feet of the normals is known in terms of the six powers λ_{ij}.

CHAPTER XIX

THE CARDIOID AND THE DELTOID

§ 137. **The Cardioid** – A tangent of the base-circle is

$$x + \bar{x}t^2 = 2t$$

The image of the point 1 in this tangent is

$$x = 2t - t^2 \tag{1}$$

This we may take as a definition of the cardioid. It has a cusp $x = 1$ when $t = 1$; and with this as base-point the equation is $x + (1 - t)^2 = 0$, or changing the sign of x

$$x = (1 - t)^2 \tag{2}$$

It is thus an inverse of the parabola with respect to the focus. Thus the focus of the cardioid is at ∞. The polarised form of (2) is

$$x = (1 - t)(1 - t_1) \tag{3}$$

This is, for given t_1, a circle on the cusp. This circle touches the curve at $t = t_1$, since in (2)

$$\frac{dx}{dt} \text{ at } t_1 = 2(t_1 - 1)$$

and in (3)

$$\frac{dx}{dt} \text{ at } t_1 = t_1 - 1$$

so that the directions are the same. It is called an osculant circle.

Two such circles, t_1 and t_2, meet at the cusp and at the point

$$x_{12} = (1 - t_1)(1 - t_2)$$

Three such circles meet at points on the line

$$x = (1 - t_1)(1 - t_2)(1 - t_3)/(1 - t) = \Pi_3/(1 - t)$$

Four such lines are included in

$$x = \Pi_4/(1-t)(1-t')$$

and are therefore tangents of the parabola

$$x = \Pi_4/(1-t)^2$$

This is merely a repetition of the argument used for the parabola.

Exercise 1 – From (2) t is the direction of the line from the cusp to x. Hence, if $x = \varrho t$,

$$\varrho t = (1-t)^2$$

or, if $t = \varrho^{\iota\theta}$,

$$\varrho = 2(1-\cos\theta)$$

§ 138. **Lines of the Cardioid** – Taking the equation § 137 (1) let us find the tangent at t. For two points t and t_1 we have

$$x = 2t - t^2$$
$$x_1 = 2t_1 - t_1^2$$
$$x_1 - x = (t_1 - t)(2 - t - t_1)$$

As t_1 approaches t along the base-circle, $t_1 - t$ has a limiting direction $\pm\iota t$, the sign depending on the side of t considered. Thus as x_1 approaches x, $x_1 - x$ also has a limiting direction, $\pm 2\iota t(1-t)$. This is the direction of the curve at x, and the line on x with this direction is defined as the tangent, or line of the curve, at x. The clinant is $\pm 2\iota t(1-t)/\mp 2\iota/t(1-1/t)$; that is, t^3 when $t \neq 1$. Thus the line t of the curve is

$$x - 2t + t^2 = t^3(\bar{x} - 2/t + 1/t^2)$$

or

(1) $$x - 3t + 3t^2 - \bar{x}t^3 = 0$$

This gives for $t = 1$ the cusp-line, $x = \bar{x}$.

For $t = -\omega$ or $-\omega^2$ it gives $x + \bar{x} = 3$. This then is a double line, tangent at both $-\omega$ and $-\omega^2$.

For a given point x, (1) is a cubic in t. When the roots are all turns there will be three tangents from x. The curve as a locus of points divides the plane into two regions; the exterior is that for which there are three tangents.

In general, the number of actual tangents depends on the regions made by the curve and its stationary lines.

Taking an exterior point x, let the roots of (1) be t_1, t_2, t_3. Then

$$s_3 = t_1 t_2 t_3 = x/\bar{x}$$

Thus the product of the clinants t_i^3 of the three tangents is the cube of the clinant of x.

Therefore, given three tangents on a point, the locus of the centre of the cardioid is made up of three lines equi-spaced—that is, making with each other the angle $\pm 2\pi 3$. In particular, if x be on the double line there is but one other tangent; then measuring angles from the double line the line to the centre trisects the angle made by the third tangent.

Exercise 2 – The tangents at t, ωt, $\omega^2 t$ are parallel.

Exercise 3 – The points at which tangents are parallel have a fixed centroid, namely, the centre.

Exercise 4 – The cusp is a Fermat point of the points of contact of parallel tangents.

Exercise 5 – Find the locus of the other Fermat point.

Exercise 6 – The longest segment of a line which can be placed in the curve joins the points ω and ω^2.

§ 139. **The Section by a Line** – The cardioid being an inverse of a parabola, the question of intersections by a circle is the same for both. The condition that four points of a parabola are on a circle was in terms of the parameter ρ,

$$\Sigma\rho_i = 0$$

or, in terms of the parameter t,

$$\Sigma(1+t_i)/(1-t_i) = 0$$

that is

(1) $$2 - s_1 + s_3 - 2s_4 = 0$$

This then is the condition that four points of the cardioid are on a circle.

It expresses that the four parameters are apolar to the cusp-parameter taken three times and the apse-parameter.

The bitangent-circles, by inversion of the results for the parabola, are when actual all outside the curve. The cusp itself, the bitangent line or double line, and the apse-circle or stationary circle are the features. Then the circles leave the curve, and tend to ∞. The imaginary circles, or extra pairs, are on the curve which is an inverse of a parabola with regard to the apse—a curve called the cissoid. This cissoid in the vertical plane has its cusp at the cusp and its focus at the apse of the cardioid.

What is proper to the cardioid is the sections by a line. Where the line $x/a+\bar{x}/\bar{a}=1$ meets the curve $x=2t-t^2$ we have

$$(2t-t^2)/a+(2/t-1/t^2)/\bar{a}=1$$

so that

$$s_1\equiv t_1+t_2+t_3+t_4=2 \tag{2}$$

Thus $s_1=2$, $s_3=2/s_4$, characterise such a section.

Exercise 7 – The sum of the distances from the cusp to the sections by a line is a constant.

If the tangents at three of these points meet at a point, then

$$\begin{aligned} 3/\bar{x} &= t_1+t_2+t_3 \\ &= 2-t_4 \end{aligned}$$

The locus of points x, the tangents from which have their points of contact on a line, is then the circle

$$3/x=2-t$$

Exercise 8 – This is the circle on the cusp and the points of contact of the double line.

Strictly, for three actual tangents, only the cusp and the exterior arc of the circle apply. The rest of the circle applies to the case when x is within the cardioid. The roots of the line-equation

$$x-3t+3t^2-\bar{x}t^3=0$$

are then t, a, $1/\bar{a}$. To these correspond a point on the cardioid, and an image-pair of it. The image-pair defines as in § 83 a tangent antilogous form (two reciprocal antilogies) whose fixed point is x. When x is on the interior arc of the above circle, the axis of the image-pair is on the point of contact of the one tangent line from x.

§ **140. Two Cardioids** – Consider two cardioids with centres 0 and c. The line-equations may be taken as

$$x-3t+3t^2-\bar{x}t^3=0 \tag{1}$$

and

(2) $$x - c - 3a\tau + 3\bar{a}\tau^2 - (\bar{x} - \bar{c})\tau^3 = 0$$

For a common line of both curves we must have $\tau^3 = t^3$; that is, $\tau = t$, ωt, or $\omega^2 t$.

Taking the possibility $\tau = t$, we must also have

(3) $$c - 3(1 - a)t + 3(1 - \bar{a})t^2 - \bar{c}t^3 = 0$$

Let all the roots be turns, t_i. Then

(4) $$c/\bar{c} = t_1 t_2 t_3 \equiv s_3$$

or, since the clinant of a line t is t^3, the cube of the clinant of the join of the centres is the product of the clinants of these three common lines. The same is true if we take $\tau = \omega^n t$, for we then have in place of (3)

$$c - 3(1 - a\omega^n)t + 3(1 - \bar{a}\omega^{2n})t^2 - \bar{c}t^3 = 0$$

Thus when all the common tangents are nine lines (that is, when no antilogies enter) these lines fall into three sets of 3.

From (4) the centres of the cardioids which touch three given lines lie on a system of lines, whose clinants are the three cube roots of the product of the three clinants of the given lines. Let us call these lines the axes of the given triangle, and suppose that there are n of them. When the centres of two cardioids are on an axis, the triangle is a set, for these two cardioids. The transition from one axis to another occurs when one of the given lines is a double line, say a_2, a_3. We know that the centre is then found by trisecting the angles a_3, a_2, a_1; a_2, a_3, a_1, so that there are nine centres when a_2, a_3 is the double line, or twenty-seven in all.

The axes then, of which $n/3$ have an angle ϕ, $n/3$ an angle $\phi + 2\pi/3$, and $n/3$ an angle $\phi - 2\pi/3$, meet at twenty-seven points. Hence $n = 9$. The axes are then nine lines, three in each of three equispaced directions.

Exercise 9 – If p_i be the distances of a point from the edges of the triangle and a_i the angles, the nine axes are three such as

$$p_1 \sin\frac{\pi - a_1}{3} + p_2 \sin\frac{\pi - a_2}{3} + p_3 \sin\frac{-\pi - a_3}{3} = 0$$

and six such as

$$p_1 \sin\frac{2\pi - a_1}{3} + p_2 \sin\frac{\pi - a_2}{3} + p_3 \sin\frac{-2\pi - a_3}{3}$$

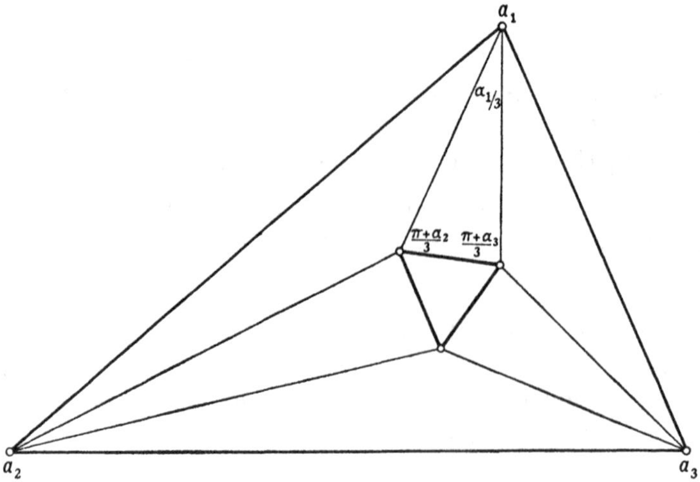

FIG. 64

Consider in particular the cardioids which lie inside the triangle. Let c_1 be the centre of that of which a_2a_3 is a double line.

Then

$$\angle\, a_3,\ a_2,\ c_1 = \alpha_2/3$$
$$\angle\, c_1,\ a_3,\ a_2 = \alpha_3/3$$

and we have seen that the lines c_1c_2, c_2c_3, c_3c_1 form an equilateral triangle.*

Exercise 10 – Verify this by trigonometry.

Exercise 11 – Verify from the figure that the lines of the triangle a_i make with any line of the triangle c_i angles whose sum is 0 (mod π).

§ **141. The Deltoid** – We have seen the convenience of treating the cardioid from the line-equation

$$x - 3t + 3t^2 - \bar{x}t^3 = 0$$

Let us now consider the curve whose line-equation is the cubic in t,

(1) $$t^3 - xt^2 + \bar{x}t - 1 = 0$$

---

* For references on this theorem see Lob and Richmond, *Proc. London Math. Soc.*, ser. 2, vol. 31.

For the analogous theory of the axes of an n-line, see *American Journal of Mathematics*, vol. 51 (1929); *Extensions of Clifford's Chain-Theorem*, by F. Morley, and a note by P. S. Wagner, *ibid.*

There is a region for which three lines can be drawn from a point. For then the discriminant of (1) is positive. We consider a point x of this region. We have then

(2) $$x = t_1 + t_2 + t_3 \equiv s_1$$

and

(3) $$1 = t_1 t_2 t_3 \equiv s_3$$

Hence two lines t_1, t_2 meet at the point

(4) $$x = t_1 + t_2 + 1/t_1 t_2$$

and in particular when $t_1 = t_2 = t$ we have the map-equation of the curve

(5) $$x = 2t + 1/t^2$$

The curve has cusps when $dx/dt = 0$; that is, when

$$t = 1, \quad \omega, \quad \omega^2$$

The cusps are then 3, 3ω, $3\omega^2$. For each of these (1) is a cube. Thus the cusp-lines are from (1)

$$x = \bar{x}$$
$$x = \omega^2 \bar{x}$$
$$x = \omega \bar{x}$$

They meet at the base-point, which is the centre of the curve. The curve has three reflexions into itself in the cusp-lines. Hence the cusp-lines meet the curve again at apses.

The curve from (5) is a hypocycloid with three cusps. We call it a deltoid.

The equation (4) is for varying t_2

$$x = t_1 + t + 1/t_1 t$$

the segment of a line. The mid-point of the segment is the point t_1 of the base-circle and the ends or cusps of the segment are given by $dx/dt = 0$; that is, $t^2 t_1 = 1$. They are then the points $x = 2t + 1/t^2$, on the curve.

Thus a rod of length 4 can be moved always touching the curve and with its ends on it (fig. 65).

The equation of the motion is

$$x = ty + 1/t^2 \tag{6}$$

The points $y = \pm 2$ mark in the x-plane the deltoid. The cusps $x = 3,\ 3\omega,\ 3\omega^2$ of the deltoid mark in the y-plane the curve

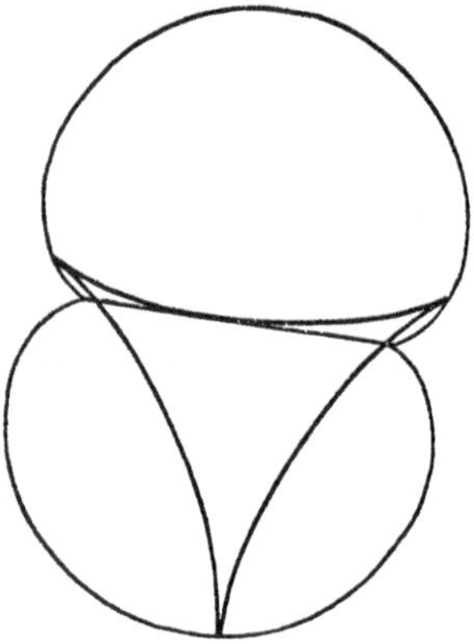

Fig. 65

$$y = 3/t - 1/t^3 \tag{7}$$

an epicycloid with two cusps. With every epicycloid there is a hypocycloid with the same cusps in the same order. If we call these complementary then (7) is the complement of the segment.

Exercise 12 – If, substituting from (7) in (6), we find the envelope of (7), we get the complement of the deltoid, as well as its cusps.

The region within the deltoid is, according to Professor Kakeya, the region of least area within which a rod, taking in its motion all possible orientations, can be reversed.*

Let us consider the deltoid of which four lines are given. We see from

$$x = t_1 + t_2 + 1/t_1t_2$$

that three lines t_i give three points

$$x_i = s_1 - t_i + t_i/s_3$$

on a circle whose centre is s_1 and radius $|\,1 - s_3\,|$. The orthocentre is $x = s_1 - (1 - 1/s_3)s_1$; that is, s_1/s_3. Hence the centre of the deltoid which touches three given lines is equidistant from the circumcentre and the orthocentre; that is, it is on the axis of these two points. Hence for four lines the four such axes meet at a point, which is the centre of the unique deltoid on the four lines.

The centre of the nine-point circle of the three lines t_i is given by

$$2x = s_1(1 + 1/s_3)$$

* See W. B. Ford, *Bulletin of the American Mathematical Society*, vol. 28 (1922), p. 45; and *American Mathematical Monthly*, vol. 29 (1922), p. 160.

Thus for 3/4 lines it is included in

$$2x = (s_1 - t)(1 + t/s_4)$$

where now s_i refers to the four t_i.

The conjugate equation is

$$2\bar{x} = (s_3/s_4 - 1/t)(1 + s_4/t)$$
$$= (s_3/t - s_4/t^2)(1 + t/s_4)$$

Hence

$$2(tx + \bar{x}) = s_2(1 + t/s_4)$$

a line on the point $x = s_2/2s_4$, and perpendicular to the line t of the curve.

Hence if from the nine-point centre of 3/4 lines a perpendicular be drawn to the fourth, the four such perpendiculars meet at a point. This is one extension to four lines of the orthocentric property.

Exercise 13 – The point is the foot of the perpendicular from the centre of the deltoid on the line of orthocentres of the four lines.

We have said that in general five lines have a Miquel circle. We must notice here an exception, that when the lines touch a deltoid the circle becomes a line.

The circumcircle of three lines is

$$x = t_1 + t_2 + t_3 - t + t/t_1t_2t_3$$

For four lines the four such circles meet. For the value $t = t_1t_2t_3t_4$ gives

$$x = s_1 - s_4$$

For five lines the five such points are included in

$$x = s_1 - t - s_5/t$$

where s_i now refers to the five t_i. And this is the equation of a segment—the Miquel segment.

For six lines the Miquel segments are included in

$$x = s_1 - t - t' - s_6/tt'$$

where s_i refers to the six t_i. And this is the penosculant form of a second deltoid, which is merely the given one displaced.

Thus the six Miquel segments for six lines of a deltoid are tangent segments of a second deltoid.*

* This reciprocity was pointed out by S. Kantor, "Die Tangentengeometrie an der Steinersche Hypocycloide," *Wiener Berichte*, vol. 78, p. 232. For further exceptions to what is known as Clifford's chain, see F. Morley, *Trans. Amer. Math. Soc.*, vol. 1 (1900), and W. B. Carver, *Amer. Journal*, vol. 42 (1920).

CHAPTER XX

CREMONA TRANSFORMATIONS

§ 142. **A Simple Illustration** – Let us illustrate a Cremona transformation by the following simple case. Suppose that the vector from x to y subtends a given angle at 0 and a given angle at 1. That is, let

(1) $$x\bar{y}=t_1\bar{x}y$$

and

(2) $$(x-1)(\bar{y}-1)=t_2(\bar{x}-1)(y-1)$$

If we eliminate $\bar{y}$, then

(3) $$t_1\bar{x}y/x-t_2(\bar{x}-1)(y-1)/(x-1)=1$$

In general x gives a definite y. But when $x=0$, let $\bar{x}/x=\tau$. Then

$$t_1\tau y-t_2(y-1)=1$$

which is a line for varying τ. And so when $x=1$, writing $(\bar{x}-1)/(x-1)=\tau$,

$$t_1y-t_2\tau(y-1)=1$$

another line.

And if $x=\infty$, we write again $\bar{x}/x=\tau$, so that

$$t_1\tau y-t_2\tau(y-1)=1$$

and y is on a circle.

In fact to make y arbitrary in (3) we must have

$$t_1\bar{x}/x-t_2(\bar{x}-1)/(x-1)=0$$

and

$$t_2(\bar{x}-1)/(x-1)=1$$

Hence, in addition to the singular points 0, 1, ∞, there is the singular point given by

$$t_2(x/t_1-1)=x-1$$

Exercise 1 – To this singular point corresponds the axis of reals.

Exercise 2 – When x describes a circle draw the curve of y, mechanically.

A Cremona transformation is then a one-to-one transformation in general, but with singular points a_i to which correspond not points but fundamental curves R_i. To the clinants around a_i correspond the points of R_i. Where two fundamental curves intersect, we have a point y with more than one correspondent x; that is, we have a singular point of the reciprocal transformation. When then the transformation is mutual (as in § 109), the fundamental curves intersect in the singular points.

§ **143. Focal Pairing** – In § 109 we generalised the equation of the secant of the base-circle

$$x + \bar{x}s_2 = s_1$$

by writing

(1) $$x + y + s_3\bar{x}\bar{y} = s_1$$

s_i referring to three points t_i. This we called focal pairing.

We wish now to remove the restriction that the points are on the base-circle.

Taking any three points a_i, consider

(2) $$\lambda_1/(x - a_1) + \lambda_2/(x - a_2) + \lambda_3/(x - a_3) = 0$$

where the λ_i are real. We have two roots, say x and y, of which each determines the other. For if x is given, there are in general two conjugate equations determining the ratios $\lambda_1 : \lambda_2 : \lambda_3$.

Instead of (2) we write the identity in z,

(3) $$(z - x)(z - y)\Sigma\lambda_i = \Sigma\lambda_1(z - a_2)(z - a_3)$$

whence, when $z = a_1$,

$$(a_1 - x)(a_1 - y)\Sigma\lambda_i = \lambda_1(a_1 - a_2)(a_1 - a_3)$$

showing the angle-property,

$$\angle\, a_2, a_1, x = \angle\, y, a_1, a_3$$

Exercise 3 – If $a_i = t_i$, then

$$(x + y)\Sigma\lambda_i = \Sigma\lambda_1(t_2 + t_3)$$
$$xy\Sigma\lambda_i = \Sigma\lambda_1 t_2 t_3$$

Deduce

$$x + y + s_3\bar{x}\bar{y} = s_1$$

We have then in (2) a mutual correspondence of two points

x and y. It has the singular points a_i, to which correspond curves. For when $x = a_1$, then $\lambda_1 = 0$, and the equation is

$$\lambda_2(x - a_3)(x - a_1) + \lambda_3(x - a_1)(x - a_2) = 0$$

so that y is $(\lambda_2 a_3 + \lambda_3 a_2)/(\lambda_2 + \lambda_3)$, any point on the join of a_2 and a_3.

But also it has the singular point ∞, for then from (2)

$$\lambda_1 + \lambda_2 + \lambda_3 = 0$$

and

$$y = \Sigma\lambda_1 a_2 a_3 / \Sigma\lambda_1 a_1$$

which is the circle on a_i.

To the points a_1, a_2, a_3, ∞ correspond then the lines of the triangle and the circumcircle, as we knew before.

Let us now apply a homography to (2).

Writing (2) with accents,

(2′) $$\Sigma\lambda_1/(x' - a_1') = 0$$

let

$$x' = \kappa/(x - a_4)$$

so that a_1', a_2', a_3', ∞ become say a_1, a_2, a_3, a_4.

Since

$$x' - a_1' = \kappa(a_1 - x)/(x - a_4)(a_1 - a_4)$$

(2′) becomes

$$\sum^{3} \lambda_1(a_1 - a_4)/(x - a_1) = 0$$

or

$$\sum^{3} \lambda_1/(x - a_1) - (\lambda_1 + \lambda_2 + \lambda_3)/(x - a_4)$$

or

(4) $$\sum^{4} \lambda_i/(x - a_i)$$

where

$$\lambda_1 + \lambda_2 + \lambda_3 + \lambda_4 = 0$$

This then is the general form. The singular points are now a_i (corresponding to $\lambda_i = 0$) and to them correspond the circles on the other three points—the fundamental curves.

The physical meaning is of interest. The effect at x of a force of strength λ_1 directed from a_1 and varying inversely as the distance is $\lambda_1/(\bar{x} - \bar{a}_1)$. There will be equilibrium for a system of such forces when

$$\Sigma\lambda_i/(\bar{x} - \bar{a}_i) = 0$$

that is when

$$\Sigma\lambda_i/(x - a_i) = 0$$

We may instance electric currents in long straight wires perpendicular to the plane.

Exercise 4 – For a point attracted to three points by forces of equal strength, varying inversely as the distance, find the two positions of equilibrium.

Exercise 5 – If four points are on a circle, the focal pairing becomes the inversion in that circle. If four points coincide, two and two, the focal pairing becomes a quadratic involution.

Let us take any three quadratics q_i, where

$$q_i = \beta_i x^2 - \gamma_i x + \delta_i = 0$$

We wish to show that for real λ_i the quadratics

$$\Sigma\lambda_i q_i = 0 \tag{5}$$

gives the above pairing. The equation maps the pairs of points x and y on the points (or lines) of a second plane. It sets up thus a two-to-one correspondence of two planes.

The equation (5) is homographic in x. We may then examine it for $x = \infty$. We have then

$$\lambda_1\beta_1 + \lambda_2\beta_2 + \lambda_3\beta_3 = 0$$

and

$$\lambda_1\bar{\beta}_1 + \lambda_2\bar{\beta}_2 + \lambda_3\bar{\beta}_3 = 0$$

giving definite ratios $\lambda_1 : \lambda_2 : \lambda_3$ and a definite y except when $\beta_1/\bar{\beta}_1 = \beta_2/\bar{\beta}_2 = \beta_3/\bar{\beta}_3$. In this latter case ∞ is a singular point. The equation (5) is then, taking β_i as real and including it in λ_i,

$$\sum^{3}\lambda_i(x^2 - \gamma_i x + \delta_i) = 0 \tag{6}$$

For $x = \infty$, $\Sigma\lambda_i = 0$, and we have

$$y\Sigma\lambda_i\gamma_i = \Sigma\lambda_i\delta_i$$

a circle.

The involution is then such that to any singular point corresponds a circle. It is the Desargues involution, or focal pairing.

To substantiate this, let us write from (6) the relations of x and y.

We have

$$\Sigma\lambda_i(x+y)=\Sigma\lambda_i\gamma_i$$
$$\Sigma\lambda_i\, xy \quad =\Sigma\lambda_i\delta_i$$

whence

$$\begin{vmatrix} x+y & \gamma_1 & \gamma_2 & \gamma_3 \\ \bar{x}+\bar{y} & \bar{\gamma}_1 & \bar{\gamma}_2 & \bar{\gamma}_3 \\ xy & \delta_1 & \delta_2 & \delta_3 \\ 1 & 1 & 1 & 1 \end{vmatrix}=0,$$

say

$$xy+a(x+y)+b(\bar{x}+\bar{y})+c=0$$

Under a translation this is

$$xy+b(\bar{x}+\bar{y})+c'=0$$

and under $x=\kappa x'$ this is

(7) $$xy+\tau(\bar{x}+\bar{y})+d=0$$

In addition to the singular point ∞, for which $y=\tau t$, there are singular points x for which (7) is a line. These are given by identifying (7) with its conjugate,

$$\tau\bar{x}\bar{y}+x+y+\bar{d}\tau=0$$

so that

$$x=1/\bar{x}=(\tau\bar{x}+d)/(x+\bar{d}\tau)$$

or

$$x^3+\bar{d}\tau x^2=\tau+dx$$

Taking these points as t_i, then $\tau=s_3$, $d=-s_2$, and (7) is

$$xy+s_3(\bar{x}+\bar{y})=s_2$$

the standard equation for Desargues pairing.

It thus appears that the two-to-one correspondence

$$\lambda_1q_1+\lambda_2q_2+\lambda_3q_3=0$$

where the λ_i are real and q_i are quadratics, is reducible to

(8) $$\sum^4\lambda_i/(x-a_i)=0$$

The four-point pairing takes a very simple form when we regard the four points a_i as on a sphere. We have then a tetrahedron of reference, and use barycentric co-ördinates. Let those of x any point be ξ_i and those of any point y be η_i. We consider then the Cremona pairing,

(9) $$\xi_i\eta_i=\kappa_i$$

But we limit this by asking that the sphere on a_i be sent into itself. The equation of the sphere is

$$(10) \qquad \Sigma\lambda_{12}\xi_1\xi_2 = 0$$

where λ_{12} is the squared distance of an edge. This becomes under (9)

$$\Sigma\lambda_{12}\kappa_1\kappa_2\eta_3\eta_4$$

and identifying this with (10),

$$\lambda_{12}\kappa_1\kappa_2 = \mu\lambda_{34}$$

Therefore

$$(11) \qquad \begin{aligned} \kappa_1 &= \lambda_{23}\lambda_{34}\lambda_{42} \\ \kappa_2 &= \lambda_{34}\lambda_{41}\lambda_{13} \\ \kappa_3 &= \lambda_{41}\lambda_{12}\lambda_{24} \\ \kappa_4 &= \lambda_{12}\lambda_{23}\lambda_{31} \end{aligned}$$

Any plane $\Sigma a_i\xi_i$ becomes $\Sigma\kappa_i a_i/\eta_i$, a cubic surface with double points at the reference points, cutting the sphere in a bicubic curve, also with double points at the reference points. Thus the pairing sends a circle into a rational bicubic curve R^3. The points which the circle has in common with the fundamental circles become the double points of R^3.

In particular, when the circle is a double point, its transform R^3 acquires a fifth double point. It then breaks up into the five double points, having conjugate factors

$$x^2\bar{x} + \ . \ . \ .$$

and

$$x\bar{x}^2 + \ . \ . \ .$$

each factor having a single point at x_i.

The image-form of R^3 has the factor

$$x^2\bar{y} + \ . \ . \ .$$

Thus we have the (2, 1) transformation

$$(12) \qquad \bar{y} = (a_0x^2 + a_1x + a_2)/(\beta_0x^2 + \beta_1x + \beta_2) \ . \ . \ .$$

In this transformation there are for a given y two points x. These points x are then in an involution. Taking the double points as 0 and ∞, we have (12) in the form

$$(13) \qquad (x^2 - a^2)(\bar{y} - \bar{\beta}) = \gamma$$

There may be five fixed points, or three fixed points and an interchange, or one fixed point and two interchanges. Consider the first case. We have then a five-point or quintic, a_i. It gives a transformation (12) or (13), and hence an involution giving a second quintic, b_i. We prove that a_5 and b_5 are a focal pair as to a_1, a_2, a_3, a_4 (or as to b_1, b_2, b_3, b_4). For this is so if the four-rowed determinant

$$| \, 1, \quad 1/(a_5 - a_1), \quad 1/(\bar{a}_5 - \bar{a}_1), \quad 1/(a_5 + a_1 \, |$$

vanishes. But since the a_i are given by

$$(x^2 - \alpha^2)(\bar{x} - \beta) = \gamma$$

we have

$$\bar{a}_5 - \bar{a}_1 = \gamma/(a_5{}^2 - \alpha^2) - \gamma/(a_1{}^2 - \alpha^2)$$

whence $1/(\bar{a}_5 - \bar{a}_1)$ is of the type $\lambda + \mu/(a_5 - a_1) + \gamma/(a_5 + a_1)$ and the determinant vanishes.

If we take one of the five points as ∞, then in (12) $\beta_0 = 0$. If we then take three of the finite points as t_i we have (12) in the form

(14) $$\bar{y} = \{a(x^2 - s_1 x + s_2) + 1\}/(x + s_3 a)$$

for this is true when $x = y = t_i$.

Let $2m$ be the sum of the roots of this quadratic in x. Then

$$\bar{y} = a(2m - s_1)$$

This is an antilogy defined by the four finite points. For each point there is a conic, with it as focus, inscribed in the triangle formed by the other three. Whereas the involution above sends each point into the other focus, the antilogy sends each point into the centre of its conic. It sends the circumcentre $y = 0$ into the nine-point centre $s_1/2$.

Exercise 6 – The four circumcentres and the four nine-point centres form negatively similar quadrangles (H. W. Richmond).

Exercise 7 – To the point $x = \infty$ in (14) corresponds in the involution the point $x = -s_3 a$. In the case when this point is outside all four circumcircles, each circle subtends the same angle at the point (G. T. Bennett).

Exercise 8 – The antilogy sends Bennett's point into a point on each nine-point circle. Thus the four nine-point circles meet at a point.

Exercise 9 – When a focal pair is outside (or inside) the four circumcircles, it has four hyperbolic distances (one for each circle). Prove that these distances are equal.

§ 144. **Intrinsic Co-ordinates** – This four-point pairing may be treated by the intrinsic co-ordinates of § 41.

We take two points θ and ϕ mutually related by

(1) $$\theta_i + \rho_i = \kappa_i, \qquad i = 1, 2, 3$$

In general a point θ gives a definite point ϕ. But when $\theta = \kappa$, $\phi_i = 0$, and ϕ is any point of the circumcircle. And when θ is a reference point x_1, so that $\theta_1 = 0$ but θ_2 is arbitrary, then $\phi_1 = \kappa_i$, the equation of a circle on x_2, κ, x_3. Thus we have the four singular points x_1, x_2, x_3, κ.

If we combine this transformation with the inversion $\bar{1}\,\bar{2}\,\bar{3}$; that is, replace ϕ_i by $-\phi_i$, then

(2) $$\theta_i - \phi_i = \kappa_i$$

This is again a Cremona transformation. It has the period n when $n\kappa_1 = 0$, $n\kappa_2 = 0$, $n\kappa_3 = 0$ (mod π). In particular it is of period 3 when

$$\theta_i - \phi_i = \pi/3$$

or

$$\theta_i - \phi_i = -\pi/3$$

that is when the fourth singular point is a Hessian point; in other words, when the four points are self-apolar. This transformation

$$\theta_i - \phi_i = \pi/3$$

sends ∞, for which $\theta_i = -a_i$, into

$$\phi_i + a_i = \pm 2\pi/3$$

that is into the points at which the edges subtend equal angles—the Fermat points, say f_1 and f_2.

Thus the given three-point x_i yields a new three-point, ∞, f_1, f_2. The relation is a mutual one.

Exercise 10 – (1) The two three-points are apolar; (2) the Hessian pair of each is in the antigraphy of the other; (3) the Jacobian of the two Hessian pairs is the intersection of the circumcircles.

§ 145. **The Geiser Transformation** – As a second case of a Cremona transformation of period 2, or Cremona pairing, consider all biquadratics C^2 on six given points a_i. Two such biquadratics have then two other common points, x and y. And all biquadratics on a_i and x meet at y. There is then a pairing of points in general. But when x is a_1, then C^2

acquires a double point at a_1; it is determined by this double point (three conditions) and the five other points a_i. Thus y is any point on this C^2, which we denote by R_1^2. Thus a_1 is a singular point, and R_1^2 is a fundamental curve. There is then for six given points a Cremona pairing T_6 with the points a_i as singular points, and the rational curves R_i^2 as fundamental curves.

When now x describes a circle, to the four points common to this circle and R_i^2 correspond the point $y=a_i$ taken four times—a fourfold point. Thus the curve which is the transform of the circle is a C^n with six fourfold points a_i. To two circles correspond two C^n's, which must therefore have two free intersections.

Thus

$$2n^2=6\times 16+2$$

or

$$n=7$$

A curve of inversive order 7 with six fourfold points, each counting as six double points, is rational. Therefore to a circle, under T_6, corresponds an R^7, with six fourfold points a_i.

But when the circle is on a singular point, a_1, then the fundamental curve R_1^2 peels off, and there is left a curve R^5 with a double point at a_1, and triple points at $a_2 \ldots a_6$, the equivalent of sixteen double points.

And when the circle is on a_1 and a_2, then R^7 becomes $R_1^2R_2^2R^3$, where R^3 has a_1, a_2 as single points and a_3, a_4, a_5, a_6 as double points. Lastly to the circle on a_1, a_2, a_3 corresponds the circle on a_4, a_5, a_6.

When the six points are taken on a sphere the biquadratics are the sections of the sphere by quadrics. When of the eight points common to three quadrics six are fixed, the mutual transformation of the other two is the Geiser transformation. We are stating this above in inversive language, with reference to a fixed quadric of the system.

When the circle becomes a double point, the R^7 acquires an extra double point, and breaks up. The factors are of the form

$$x^4\bar{x}^3+\ldots$$

and

$$x^3\bar{x}^4+\ldots$$

each having double points at a_i.

§ **146. The Bertini Transformation** – The circles of the plane (1) have two common points, and (2) are a triply infinite linear system.

Let us ask what other system of rational curves C^n has these properties. There will be multiple points a_i. We suppose all of the same multiplicity μ, that is when such a point is 0 all the terms $x_\alpha \bar{x}_\beta$ for which $\alpha + \beta < \mu$ are absent. Thus for $\mu = 1$ the point a_i is on the curve; for $\mu = 2$, a_i is a double point; for $\mu = 3$ it is a triple point, and so on. Let there be λ points a_i, each of multiplicity μ.

Then since two curves C^n meet in $2n^2$ points,

$$\lambda\mu^2 = 2(n^2 - 1) \tag{1}$$

and since a μ-fold point imposes $\mu(\mu+1)/2$ linear conditions,

$$\lambda\mu(\mu+1)/2 = (n+1)^2 - 4 = (n-1)(n+3) \tag{2}$$

Hence

$$\lambda\mu(\mu-1)/2 = (n-1)^2 \tag{3}$$

which is the condition that the curves be rational.

Also

$$(n+1)/(n-1) = \mu/(\mu-1)$$

that is

$$n = 2\mu - 1 \tag{4}$$

$$\lambda = 8 - 8/\mu \tag{5}$$

Hence we have the table

$\mu = 1$	2	4	8
$\lambda = 0$	4	6	7
$n = 1$	3	7	15

That is, we have

(1) all circles;
(2) curves R^3 with four given double points;
(3) curves R^7 with six given quadruple points;
(4) curves R^{15} with seven given eightfold points.

We have considered the case (2) and briefly the case (3). The case (4) is that known in projective geometry as the Bertini transformation (Pascal, *Repertorium*, vol. 2, p. 370).

In the Bertini transformation T_7 from the inversive view we have seven points a_i; two curves C^4 with these as double points have $2 \times 16 - 7 \times 4$ or four intersections, but these pair off. A curve C^2 on the points and a curve C^4 twice on the points have two other intersections, which are a pair of T_7.

CHAPTER XXI

THE n-LINE

§ 147. **Cyclogens** – We have indicated in Chapter XV the advantage of referring a triangle to its circumcircle. And we have seen the simplicity of regarding four lines as lines of a parabola or as osculant lines of a cardioid. The cardioid was strictly analogous to the circle, the equation of the curves being

$$x - 2t + \bar{x}t^2$$
$$x - 3t + 3t^2 - \bar{x}t^3$$

and the lines being

$$x - s_1 + \bar{x}s_2 = 0$$
$$x - s_1 + s_2 - \bar{x}s_3 = 0$$

in the first case for $2/3\ t_i$, in the second for $3/4\ t_i$.

It is natural, then, to study the n-line from this same point of view generalised.

A curve given by $x = f(t)$ has a cusp when $dx/dt = 0$. Let then

(1) $$dx/dt = \kappa(t - \tau_1)\ .\ .\ .\ (t - \tau_{n-2})$$

or

(2) $$dx/dt + (n-1)\{a_1 + (n-2)a_2 t + \ .\ .\ .\ + \bar{a}_1 t^{n-2}\} = 0$$

Then we have defined a curve whose map-equation is

(3) $$x - x_0 + (n-1)a_1 t + \binom{n-1}{2} a_2 t^2 + \ .\ .\ .\ + \bar{a}_1 t^{n-1} = 0$$

whence the line-equation is

(4) $$x - x_0 + n a_1 t + \binom{n}{2} a_2 t^2 + \ .\ .\ .\ + n\bar{a}_1 t^{n-1} + (\bar{x} - \bar{x}_0)t^n = 0$$

for the envelope of this line is obtained by differentiation as to t, and the derivative is $n\ \times$ the conjugate of (3).

We call x_0, the value of x when $t = 0$, the centre of the curve, and take it to be 0. We have then in (3) or (4) a convenient curve as a base for the discussion of $n+1$ lines. It was indicated in § 87, where we began with the lines, and obtained such a curve uniquely.

If we write a self-conjugate polynomial in t and regard the end-coefficients as variable, we have the curve. In the abridged notation it is

$$(at)^n = 0 \tag{4}$$

when $(at) = a_0 + a_1 t$, $a_0{}^r a_1{}^{n-r}$ is a coefficient a_r, conjugate to $a_0{}^{n-r} a_1{}^r$, and $a_0{}^n = x$, $a_1{}^n = \bar{x}$.

We call such a curve a cyclogen of aspect n, n being the number of parallel tangents. And we denote it by B^n.*

When all coefficients a_i except two vanish the curve is an epicycloid, and when all vanish it is a point (taken n times).

The equation (3), with $x_0 = 0$, is in the abridged notation

$$a_0(at)^{n-1} = 0 \tag{5}$$

This is the map-equation of B^n.

The equation (2) is

$$a_0 a_1 (at)^{n-2} = 0 \tag{6}$$

a self-conjugate equation giving the cusp-parameters τ_i. To every root which is a turn corresponds an actual cusp; there are also in general pairs of roots, such as z and $1/\bar{z}$, giving elliptic pairs of cusps.

The cusps themselves are given by (5) and (6), so that

$$a_0{}^2(at)^{n-2} = 0$$

Save as to homologies $x = ay + b$, the curve is defined by its cusp-parameters. Under the group of reflexions it has $n+1$ parameters. The form $(at)^n$ has only $n-1$ parameters—the number of fixed coefficients a_r and $\bar{a}_r$, but here the base-point is taken at the centre. Thus the circle, B^2, has three parameters; the cardioid, B^3, has four, for example two for the centre and two for the cusp.

The cyclogen B^1 is a point. Here $dx/dt = 0$ is the original

* The curve B^5 is discussed by Father E. C. Phillips, *On the Pentacardioid* (The Lord Baltimore Press, 1909). We may of course, as in the deltoid, regard other than the end coefficients as variable. The curve so derived from a quintic with a_2 and a_3 as conjugate variables, is discussed by R. P. Stephens, *On the Pentadeltoid*, *Trans. American Math. Soc.*, vol. 7 (1906).

equation. It is convenient here to regard a line as a cyclogen of aspect 0, B°. If $dx/dt=\kappa/t$, and $t=e^{\iota\theta}$, then $x=x_0+\iota\kappa\theta$, a line.

When we associate with an $(n+1)$-line its B^n, the coefficients a_i of the B^n are constants of the lines under displacements. They are invariants of the lines under homologies. They form with the clinants of the line a complete set of invariants under homologies.

The B^n is a rational curve with $n-2$ marked points—the cusps. The cusp-parameters are subject to homographies which leave (6) unaltered, and thus the comitants of (6) will tell, in part at least, the story of the lines.

In a specific case we may take a_1 to be 1, and replace t by $-t$. Thus the curves, so reduced, are

$$\begin{aligned}
n&=2, &&x-2t+\bar{x}t^2, \text{ a circle}\\
n&=3, &&x-3t+3t^2-\bar{x}t^3, \text{ a cardioid}\\
n&=4, &&x-4t+6\mu t^2-4t^3+\bar{x}t^4\\
n&=5, &&x-5t+10at^2-10\bar{a}t^3+5t^4-\bar{x}t^5
\end{aligned}$$

§ **148. Osculants** – To the curve $(at)^n$ we now apply Study's theory of osculants (Study, *Leipzig Berichte*, 1886).

The polar form $(at_1)(at)^{n-1}$ is for given t the point t_1 of the curve. But for given t_1 it is a curve B^{n-1}, which is called a first osculant of B^n. (The phrase polar of a curve is preempted). This osculant touches B^n at t_1. Its map-equation is

$$a_0(at_1)(at)^{n-2}=0 \tag{1}$$

This is satisfied when

$$a_0^2(at)^{n-2}=0, \qquad a_0a_1(at)^{n-2}=0$$

Thus a first osculant is on the cusps of B^n, and also touches the curve. These are $n-1$ conditions.

Exercise 1 – The Hessian of (1) as to t is the self-conjugate equation of B_1^{n-1}.

A first osculant of B_1^{n-1}, say B_{12}^{n-2}, is a second osculant of B^n. It is, as a line equation

$$(at_1)(at_2)(at)^{n-2}$$

And so on till we come to the completely polarised form

$$(\alpha t_1)(\alpha t_2) \; . \; . \; . \; (\alpha t_n)$$

which is an osculant line. Explicitly it is

$$x + a_1 s_1 + a_2 s_2 + \; . \; . \; . \; + \bar{x} s_n = 0$$

where the s_i are product-sums of the t_i.

For $n+1$ points t_i we have $n+1$ osculant lines, and it is this form which we may take as canonical for the lines, precisely as we took for three lines the form

$$x - s_1 + \bar{x} s_2 = 0$$

for $2/3\ t_i$. It will tend to clearness if we give an account of the theory for B^3 and for B^4.

The cardioid $(\alpha t)^3$ is, reduced,

$$x - 3t + 3t^2 - \bar{x} t^3 = 0$$

The polar $(\alpha t)^2(\alpha t_1)$ is

$$x - 2t - t_1 + t^2 + 2tt_1 - \bar{x} t^2 t_1 = 0$$

This is, for given t, the point t of the curve

$$x = 2t - t^2$$

For given t_1 it is the line-equation of the osculant circle at

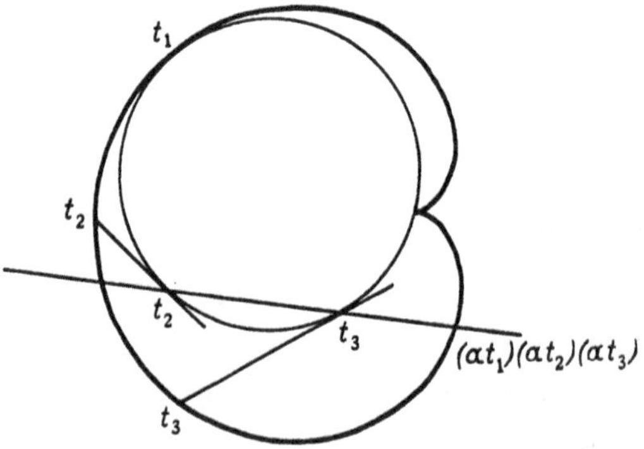

FIG. 66

t_1; a line on the point t of $(\alpha t)^3$, and making with the tangent at t the clinant t_1/t.

The two curves, the cardioid and an osculant circle, are said to be perspective; a tangent of the circle at its point t

being on the point t of the cardioid. The tangents from the latter point to the circle fall apart; that which touches at t we may call the isolated tangent.

The isolated tangents from a point t of the cardioid to two osculant circles $(at)^2(at_1)$ and $(at)^2(at_2)$ have the fixed clinant t_1/t_2. Thus as t describes $(at)^3$, the lines of a rigid pencil on t describe the osculant circles.

To obtain the envelope of osculant circles, we take the map-equation

$$x = t + t_1 - tt_1$$

By the rule of envelopes,

$$tD_t x / t_1 D_{t_1} x$$

is real—that is, $t = t_1$ or either $t = 1$ or $t_1 = 1$.

An osculant circle then touches the curve and is on the cusp. For when $t = 1$, $x = 1$, the cusp. When t_1 is 1, the osculant circle is the cusp (taken twice).

Two osculant circles $(at)^2(at_1)$ and $(at)^2(at_2)$ meet then at the cusp, and also at the point $(at_1)(at_2)(at)$. Three osculant circles meet at the cusp, and points on the line $(at_1)(at_2)(at_3)$. This line is the secant or osculant line at t_2, t_3 of the circle $(at)^2(at_1)$.

The theory of a four-line from this point of view is that the circumcircles of 3/4 lines are osculants of a cardioid. The cardioid and four points t_i on it determine the lines as $L_4 = (at_1)(at_2)(at_3)$, . . .

The intersection of the lines L_3 and L_4 is $(at_1)(at_2)(at)$. The circumcircle of $L_2L_3L_4$ is $(at_1)(at)^2$.

The curve B^4, reduced, is

$$x - 4t + 6\mu t^2 - 4t^3 + \bar{x}t^4 = 0$$

Completely polarised this gives the line

$$(2) \qquad x - s_1 + \mu s_2 - s_3 + \bar{x}s_4 = 0$$

and five lines L_i may be taken in this way, for 4/5 t_i (compare § 87).

For varying t_4 we have the intersection of L_4 and L_5, $(at_1)(at_2)(at_3)(at)$. For $t_3 = t_4 = t$, we have the circumcircle of L_3, L_4, L_5, $(at_1)(at_2)(at)^2$. For $t_2 = t_3 = t_4 = t$, we have the cardioid $(at_1)(at)^3$. The five cardioids of the 4/5 lines are osculants of $(at)^4$.

The theorem of Miquel is now that the cusps of these cardioids lie on a circle. Let us then find the cusps of the osculant cardioids. Let us here write $(at)^4$ as a map-equation,

$$x = 3t - 3\mu t^2 + t^3$$

The osculant cardioids are

$$x = 2t + t_1 - \mu(t^2 + 2tt_1) + t^2 t_1 \tag{3}$$

For a cusp, $dx/dt = 0$, that is

$$1 - \mu(t + t_1) + tt_1 = 0 \tag{4}$$

The cusp-locus is then

$$x = t + t_1 - \mu tt_1$$

or, eliminating $t + t_1$ and writing τ for tt, it is

$$\mu x = 1 + (1 - \mu^2)\tau \tag{5}$$

The cusp locus is then a circle or the arc of a circle as $|\mu| >$ or < 1.

The double points of (3)

$$1 - 2\mu t + t^2 = 0$$

give the cusp-parameters of $(at)^4$. The cusps are on the curve when $|\mu| < 1$.

In this case (5) is an arc connecting the cusps. We may call it the Miquel arc, M. It is the Miquel arc or circle of the five osculant lines formed from any 5 t_i, taken four at a time.

Exercise 2 – The tangents of the arc at the cusps are the cusp tangents of $(at)^4$.

The equation (3) is for fixed t and varying t_1, the map-equation of the osculant circle $(at)^2(at_1)^2$. Thus the envelope of these circles is the same as that of osculant cardioids $(at)^3(at_1)$. But for the latter the cusps of $(at)^4$ and $(at)^4$ itself are the envelopes, and M is the cusp-locus. For the former M and $(at)^4$ are the envelope, and the cusps are the special osculant circles which are double points.

Let us illustrate, in this case of five lines, the use of the constants a_i.

First let $a_1=0$. The centric circle of § 87 is then a point. And the circumcenters of the lines taken three at a time are all on a circle.

We place a ring of five circles (fig. 67) with centres on a given circle and each intersecting the next on the circle.

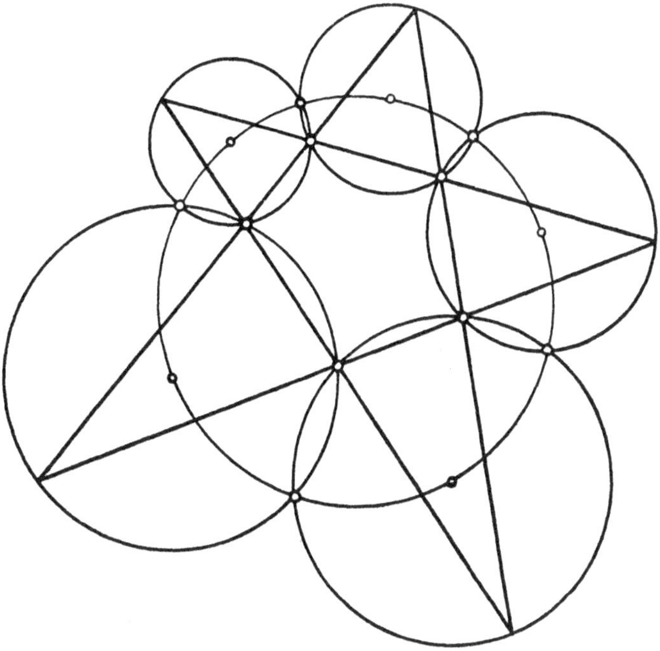

FIG. 67

The five other intersections of the adjacent circles being joined in order form the five-line, and the salient thing is that the intersection of non-adjacent sides are also on the respective five circles.

Next let $a_2=0$. The Miquel arc in this case is a segment, and (§ 141) the five lines touch a deltoid.

Again, if $a_1a_3=a_2^2$, or if in the reduced form $\mu=1$, the Miquel arc becomes a point. The five lines then touch a parabola.

Lastly five lines in general touch either an ellipse or a hyperbola according as $|\mu| >$ or < 1; that is, according as the cusps of B^4 are not or are actual.

The Miquel figure, or the complete five-line, can be constructed without having to draw circles on three points.

For, taking B^4 as

$$x = 3t - 3\mu t^2 + t^3$$

the five lines L_i meet at ten points such as

$$x_{45} = t_1 + t_2 + t_3 - \mu(t_2t_3 + t_3t_1 + t_1t_2) + t_1t_2t_3$$

The circumcentre of L_3, L_4, L_5 is then

$$x = t_1 + t_2 - \mu t_1t_2$$

The centric circle, C_1, of L_2, L_3, L_4, L_5 is

$$x = t_1 + t - \mu t_1 t$$

It is on the point μ. The centric circle of all the lines is the base-circle.

The Miquel circle M has the centre $1/\mu$ and radius $1/\mu - \mu$. It is then the circle on μ with centre at the image of μ.

Taking now a point t_1 on the base-circle, the circle C_1,

$$x = t_1 + t - \mu t_1 t$$

is the circle on μ with centre t_1. It meets M again when

$$1 - \mu t_1 t + t_1 t = 0$$

Taking two points t_1, t_2 on the base circle, and drawing circles C_1, C_2 on μ with these centres, they meet again at

$$x = t_1 + t_2 - \mu t_1 t_2$$

The circle

$$x = t_1 + t_2 + t - \mu(t_1t_2 + t_1t + t_2t) + t_1t_2t_3$$

has this as centre, and meets M where C_1 and C_2 meet it.

Three circles C_1, C_2, C_3 meet at the point x_{45}. And for four circles C_1, C_2, C_3, C_4 the four such points are on the line L_5.

§ 149. **Construction of a Cyclogen** – To construct the curve B^4 we may take the line-equation

$$x - 4t + 6\mu t^2 - 4t^3 + \bar{x}t^4 = 0$$

All these curves for varying μ are parallel. When $\mu = 0$ we have the epicycloid

$$x = 3t + t^3$$

known as the nephroid. The curve is then a parallel of the

nephroid, or an involute of its evolute, which is another nephroid.

Exercise 3 – When parallel rays of light are reflected at a circle, the envelope or caustic is a nephroid.

For the construction of a B^{n+1}, which is a special case of

$$x + a_1 t + a_2 t^2 + \ . \ . \ . \ + a_n t^n = 0$$

the method of slots and pins (§ 127) is proper.

We may also use the idea of a rigid pencil.

A first osculant of $(at)^n$, say $(at_1)(at)^{n-1}$, is perspective with $(at)^n$; that is, the tangent at t of the osculant is on the point t of the curve. The tangent of the osculant makes with the curve at t the clinant t_1/t, for the former is

$$x + \ . \ . \ . \ + \bar{x} t^{n-1} t_1 = 0$$

and the latter is

$$x + \ . \ . \ . \ + \bar{x} t^n$$

The tangents from a point of the curve to the osculant fall apart; one of them is isolated. The isolated tangents from the point t of the curve to all first osculants form a rigid pencil. For the ratio of the clinant t_1/t and t_2/t is constant.

Thus there is associated with a cyclogen a definite motion, that of the rigid pencil of tangents to first osculants in a plane y, as the vertex describes the curve in the plane x. To find the equation of the motion we have first the proper parameter τ, the relative direction of a line of the pencil to the base-line in the x-plane.

Thus $\tau^2 = t^n t_1/t$ or, taking t_1 to be -1,

$$\tau^2 = t^{n-1}$$

Second we have the centrode in the x-plane. The normals at t to the curves $(at)^{n-1}(at_i)$ meet at the point of no velocity x_0. We have then

$$x_0 = f'(\tau)$$

and by integration we have the function f, and the equation of the motion

$$y = \tau x - f(\tau)$$

Exercise 4 – The equation of the motion for B^4 is

$$yt^3 + xt^6 - 3t^4 + 3\mu t^2 - 1 = 0.$$

The point $y = 0$ describes the B^4, and any line $y = \varrho t_0$ in the plane of y envelopes a cardioid.

This construction of a cyclogen may be put thus. If we take two circles meeting at 0, a line on 0 meets them again in two points, the tangents at which meet on a B^3 with cusp at 0. If we take three circles on 0, a line on 0 meets them at three points, giving three cardioids which are osculants of a B^4. For four circles on 0, we have four B^4's which are osculants of a B^5. And so on.

Exercise 5 – It is clear that two first osculants of B^n have a common first osculant, say

$$(at)^{n-2}(at_1)(at_2)$$

Prove that if two curves B^{n-1} have a common first osculant, they are first osculants of a B^n.

§ 150. **Clifford's Chain** – The theorem of Clifford is as follows. Two lines, say 1 and 2, have a common point 12. Three lines 1, 2, 3 have three common points 12, 23, 31 which lie on a circle 123. Four lines have four circumcircles 123 which are on a point 1234. Five lines give five such points which lie on a circle 12345. And so on. We thus get the complete n-line, the lines $1 . 2 \ldots n$

$$\text{meeting in } \binom{n}{2} \text{ points } 12, \ldots$$

$$\text{these lying on } \binom{n}{3} \text{ circles } 123, \ldots$$

$$\text{these meeting in } \binom{n}{4} \text{ points } 1234, \ldots$$

ending with a point or a circle $123 \ldots n$ as n is even or odd. Regarding the lines as circles on the point ∞, we have the Clifford configuration Γ_n of 2^{n-1} points and 2^{n-1} circles, each point on n circles and each circle on n points.

To prove this we may begin with four lines. They are osculants of a cardioid $(at)^3$, which has a cusp given by

$$a_0^2(at)=0$$
$$a_0a_1(at)=0$$

Five lines are osculants of a B^4, $(at)^4$. Taken four at a time they are osculants of $(at)^3(at_i)$. The cusps of these cardioids are given by

$$a_0^2(at)(at_i)=0$$
$$a_0a_1(at)(at_i)=0$$

which give on elimination of $t+t_i$ the Miquel arc. Six lines give six B[4]'s which are osculants of a B^5, $(at)^5$. The Miquel arcs are included in

$$a_0{}^2(at)(at_1)(at_2)$$
$$a_0a_1(at)(at_1)(at_2)$$

They are then on the point given by

$$a_0{}^3(at_1)(at_2)$$
$$a_0{}^2a_1(at_1)(at_2)$$
$$a_0a_1{}^2(at_1)(at_2)$$

the Clifford point of the six-line. The point is known in advance. For the three cusp-parameters $a_0a_1(at)^3$ have an apolar pair for which

$$a_0{}^2a_1(at_1)(at_2)=0$$
$$a_0a_1{}^2(at_1)(at_2)=0$$

and the osculant of this pair is a cardioid

$$(at)^3(at_1)(at_2)$$

which is a repeated point, namely,

$$a_0{}^3(at_1)(at_2)$$

A B^{2m+1} has $2m-1$ cusp-parameters τ_i. These have a canonizant, m parameters apolar to them. These m points give an osculant B^{m+1} which is a repeated point—the Clifford point. For the cusp-parameters of any osculant of $(at)^n$, say $(at)^{n-r}(at_1)\ .\ .\ .\ (at_r)$, are the polar as to the cusp-parameters τ_i of the points $t_1\ .\ .\ .t_r$. Thus, when the polar is arbitrary, the cusp-parameters of the osculant are arbitrary. That is, for $(at)^{2m+1}$,

$$a_0{}^m a_1(at_1)(at_2)\ .\ .\ .\ (at_m)=0$$
$$.\quad .\quad .\quad .\quad .\quad .\quad .\quad .\quad .\quad .$$
$$a_0a_1{}^m(at_1)(at_2)\ .\ .\ .\ (at_m)=0$$

Whence the osculant $(at)^{m+1}(at_1)\ .\ .\ .\ (at_m)$ becomes

$$a_0{}^{m+1}(at_1)(at_2)\ .\ .\ .\ (at_m)=0$$

By eliminating the symmetric functions s_i of the t_i we have the Clifford point.

For $(at)^{2m+2}$ each first osculant has a Clifford point. The equations have the extra factor (at_{m+1}). If we eliminate all symmetric functions except the turn s_{m+1} we have a circle. This is the Clifford circle of the $(at)^{2m+2}$.

Here the simplest covariant of the $2m$ cusp-parameters τ_i is a pencil of sets of $m+1$ points apolar to τ_i. The osculant of such a set is a repeated point.

§ **151. The n-fold Parabola** – For an even number of lines we might have used, instead of the cyclogen, Clifford's n-fold parabola. This is a curve

$$x = \sum^{n} \frac{A_i}{(a_i - t)^2} \tag{1}$$

where, taking the standard case, a_i is a turn, A_i a constant, which satisfies the cusp-condition

$$\sum \frac{A_i}{(a_i - t)^3} = 0 \tag{2}$$

that is, where (2) is self-conjugate, so that it implies

$$\Sigma \bar{A}_i a_i^3/(a_i - t)^3 = 0$$

or

$$A_i = \bar{A}_i a_i^3$$

The tangent t_1 is then

$$x = \Sigma A_i/(a_i - t)(a_i - t_1)$$

this being a segment.

Two tangents t_1 and t_2 meet at

$$x = \Sigma A_i/(a_i - t_1)(a_i - t_2)$$

The circumcircle of three tangents is then

$$x = \Sigma A_i(a_i - t)/(a_i - t_1)(a_i - t_2)(a_i - t_3)$$

The cardioid of four tangents is then

$$x = \Sigma A_i(a_i - t)^2/\Pi_i^4$$

where

$$\Pi_i^4 = (a_i - t_1)(a_i - t_2)(a_i - t_3)(a_i - t_4)$$

and in general the cyclogen of m tangents is

$$x = \Sigma A_i(a_i - t)^{m-2}/\Pi_i^m$$

The number of lines which can be taken without restriction is $2(n+1)$, and when m is $2(n+1)$ the cyclogen is

$$x = \sum^{n} A_i(a_i - t)^{2n}/\Pi_i^{2(n+1)}$$

This is what we should obtain by writing the cusp-equation for B^{2n+1} in the canonical form

$$\sum^{n} a_i(t-\tau_i)^{2n-1}=0$$

giving

$$x=\sum^{n} a_i(t-\tau_i)^{2n}$$

as the canonical form of B^{2n+1}, the base-point being the Clifford point.

The theorem of Homersham Cox in the *Quarterly Journal*, vol. 25, p. 67, is so closely connected with that of Clifford that we state it. In a Euclidean space take a point p, and on it three planes π_i. On each pair of planes take a point. On these points p_{23}, p_{31}, p_{12} take a plane π_{123}. We have so far merely constructed a tetrahedron.

But on p take four planes π_i and on each pair $\pi_1\pi_2$ a point p_{12}. We get four planes π_{123}. These meet in a point p_{1234}.

Again on p take five planes π_i. There are five points p_{2345}. These are on a plane π_{12345}.

And so on *ad infinitum.*

This Cox configuration becomes the Clifford configuration when the points are all on a sphere.

§ **152. Foci** – Clifford (*Works,* p. 51) proves the theorem first by the consideration of foci. The method does not easily lend itself to the special cases where circles become lines. We had such a case in § 141. For other special cases see W. B. Carver, *American Journal,* vol. 42 (1920), pp. 137–167. But the method is highly suggestive, and we push it a little further.

We are considering a curve given by a homogeneous line-equation $(a\rho)^n=0$, where ρ_1, ρ_2, ρ_3 are distances to a line from three reference-points. The foci (§ 118) are $(ax)^n=0$, where x_1, x_2, x_3 are vectors to a focus from the reference points. To touch a given line is one linear condition on the coefficients of the curve; to have a given focus is two linear conditions; but when the given focus is ∞ it gives only one linear condition.

Consider the case $n=3$. Let there be six given tangents and a focus at ∞. There are two other foci x and y. Each in general determines the other, for we have nine conditions when x is given.

The curves $(a_i\rho)^3$ which touch six lines are included in $\sum^4 \lambda_i(a_i\rho)^3$, where the λ_i are real. Their foci are then given by $\sum^4 \lambda_i(a_i x)^3$, where $(a_i x)^3$ is a cubic in x whose leading coefficient is real, and may be taken as 1. A focus is at ∞ when $\Sigma\lambda_i = 0$. Thus the finite foci are given by

$$\sum^3 \lambda_i(\beta_i x^2 - \gamma_i x + \delta_i) = 0$$

that is (§ 143), by what in anticipation we called focal pairing.

In the case of six lines, when x is the point 12, then y is the focus of the parabola on 3456. And when $x = \infty$, y is the Clifford point 123456. Thus the transformation interchanges the points of the Clifford configuration Γ_6. But the circles of the configuration are sent into bicubic curves with the singular points as double points.

Exercise 6 – The three foci x, y, z of line-cubics $(a\rho)^3$ on five lines form an involution I_2^1 such that any two in general determine the third. The neutral pair are the foci of the conic on the lines. If this conic be

$$2x = t + \mu^2/t$$

and the five lines be t_i, prove that

$$\begin{aligned} &2\mu(\bar{x}+\mu)(\bar{y}+\mu)(\bar{z}+\mu)/\Pi_5(\mu t_i + 1) \\ &= (\mu^2+1)(x+\mu)(y+\mu)(z+\mu)/\Pi_5(t_i+\mu) \\ &+ (\mu^2-1)(x-\mu)(y-\mu)(z-\mu)/\Pi_5(t_i-\mu) \end{aligned}$$

This and its conjugate define the involution. By taking $z = \infty$, $= \bar{z}\tau$, we have the transformation which sends Γ_5, the complete five-line, into itself. It is a contact-transformation.

INDEX

The References are to the Sections